BUILDING THE WOODEN WALLS

A detail from
'A View of a Shipyard on the Thames'
painted by John Cleveley the Elder in 1762.
(Glasgow Museums and Art Galleries)

BUILDING THE WOODEN WALLS

The Design and Construction of the 74-Gun Ship Valiant

BRIAN LAVERY

Naval Institute Press

First published in Great Britain 1991 by
Conway Maritime Press Ltd,
24 Bride Lane, Fleet Street,
London EC4Y 8DR

Published and distributed in the United States of America
and Canada by the Naval Institute Press,
Annapolis, Maryland 21402

Library of Congress Catalog Card No. 91–60206

ISBN 1–55750–078–9

Printed in Great Britain

Contents

Introduction

The 'Wooden walls of England' were the ships of her navy. Other countries relied on armies and stone walls round their cities for their defence, whereas Britain relied mainly on sea power. The expression is an old one; in 1635, Lord Coventry wrote that 'The wooden walls are the best walls of this kingdom'. One hundred and eighty years later, it was still used to describe the great fleets of Nelson, though the aggressive tactics of that admiral were very different from the passive defence implied in the phrase 'wooden walls'. The expression conjures up an image of great fleets at sea, but this book looks at the wooden walls from another angle; from the point of view of the men who designed and built them – the dockyard craftsmen, the ship designers and the managers of the shipbuilding yards of the eighteenth century. The names of these men are recorded in full in the dockyard muster books in the Public Record Office, but we know little of their individual characters, or their daily lives outside working hours. This book attempts to record their skills and practices, and to discover how a ship was put together.

This book is the result of work on the Wooden Walls exhibition in Chatham Historic Dockyard. Like the exhibition, it tells the story of the building of the 74-gun ship of the line *Valiant* during the Seven Years War more than two centuries ago. Unlike the exhibition, it does not attempt to personalise or dramatise the story, but concentrates on the bare facts of shipbuilding during the period. Unlike other publications such as Peter Goodwin's *Construction and Fitting of the Sailing Man of War*, Longridge's *The Anatomy of Nelson's Ships*, and the Anatomy of the Ship series, this book looks at the actual process of shipbuilding, rather than the details of a completed ship. It is concerned with the techniques of the draughtsman, sawyer, shipwright or caulker, as much as with the product he worked on.

The *Valiant* was one of the first and most successful of the British 74-gun ships; she was built at Chatham, and was probably one of the first ships to be 'laid off' in the new mast houses and mould loft of 1753. She was copied from the French *Invincible*, and this is particularly relevent for two reasons. I have worked on the artefacts recovered from that ship after her loss in 1758, and written the standard book on it. Furthermore, the best of the artefacts were acquired by Chatham Historic Dockyard, and a selection of these is now on display at the Wooden Walls exhibition.

The *Valiant* is taken as an example, but information on the actual building of the *Valiant* is rather limited. Her original plans are available, the records of her ordering survive, and it is easy to trace the genesis of her design. Her operational career can be traced in great detail through log books, muster books, letters and other documents. But, as with all British warships of the eighteenth century, detailed information on the actual building does not exist. Dockyards treated the building of a warship, even a rather special one such as the *Valiant*, as a matter of routine and only corresponded with the Admiralty or the Navy Board about it if serious difficulties arose. Detailed records were certainly kept of the week to week work on each vessel, but unfortunately only two volumes of these have survived. Both relate to Portsmouth Dockyard, and neither covers the period under discussion in this book. As a result, we are left to interpret the actual sequence of building of the *Valiant* from other sources and other ships. These sources, including textbooks, manuscripts, draughts, paintings and models, are enough to give an accurate and complete picture, though only in general terms.

This approach has certain advantages. Though the *Valiant* is taken as the sample ship, and Chatham Dockyard as a typical building yard, most of the information has a much wider application. Construction techniques did not change dramatically between 1715 (when the Navy Board laid down new rules affecting the frames, wales and structure of ships), and 1811, when Sir Robert Seppings introduced his system of diagonal bracing. The book deals specifically with ships of the line, but the manner of building frigates and sloops was not radically different; thus the book covers the hundreds of British warships built throughout the eighteenth century, from the age of Queen Anne to the age of Nelson.

I would like to thank the staff of Chatham Historic Dockyard for their support in this work; in particular the chairman, Sir Steuart Pringle, and the curator, Richard Holdsworth. Thanks are also due to the staffs of the Public Record Office at Kew, the National Maritime Museum at Greenwich, the Science Museum in London, the British Library and the London Library.

BRIAN LAVERY, Chatham, 1991

1

The Ship of the Line

For nearly two hundred years, the ship of the line was the dominant weapon on the seas. It was built of wood, propelled by wind alone and it fired solid shot from smooth bored, muzzle-loading guns. Unlike its predecessor, the galley, it was built to withstand all kinds of weather. Unlike its successor, the steam battleship, it was not subject to attack from submarine, mine, torpedo or aircraft. The ship of the line was vulnerable to the wind and weather, and many were lost through the effects of the sea; but of forces under human control, a ship of the line could only be destroyed or taken by another ship of the line or fireships in very exceptional and rare circumstances. Smaller warships, such as frigates, did not have the firepower to penetrate the oak sides of a ship of the line, while their own sides were easily destroyed by the larger ship's great broadside.

The Line of Battle

The ship of the line, as its name implies, was intended to form part of the 'line of battle'. A ship is naturally longer than it is broad, and in this period it had to carry a large number of guns on each side to give it significant gun power. To deploy this power, a fleet was arranged in a single line ahead so that the guns of one ship did not mask those of another. This tactic was almost certainly first used by the English fleet at the Battle of the Gabbard against the Dutch in 1653; it resulted in a decisive victory, so the line of battle soon became standard with the English, and was quickly adopted by other navies.

The First Ships of the Line

One consequence of the line of battle was that ships had to be selected to form part of it. In the older form of battle, several small ships could easily combine against a large one; but the line of battle, in its most pure form, demanded that each ship keep its station against its opponent in the enemy line, however large it might be. Therefore, ships that were too small, weakly gunned or lightly constructed could not be allowed to form part of the line. The minimum size for a ship of the line tended to increase over the years from about 30 guns in the 1650s to 50 guns by the 1660s and 60 guns by the middle of the next century. However, it was generally agreed that a ship of the line needed at least two full decks of guns, while the largest ones, often used as fleet and squadron flagships, had three decks.

The English navy of the seventeenth century was first to develop the line of battle, and even before that it had gone some way to producing ships which were capable of fighting in line. The first real three-decker, *Sovereign of the Seas*, had been built at Woolwich in 1637 and she was to play an important part in many fleet battles until her loss in 1697. The ancestor of all the two-decked ships of the line was the *Speaker* of 1650, developed from a type of fast 'frigate' copied from those built at Dunkirk, but soon expanded and built up to make a very different kind of ship. The *Speaker* too had a distinguished career (mostly under the name of *Mary*) and fought in no less than fourteen battles, before being lost in 1703.[1] Both the *Sovereign* and the *Speaker* types were further developed between 1660 and 1685, notably in Samuel Pepys's great 'Thirty Ships' programme of 1677. As a result, there were three major types in the line of battle by the end of the period: large three-deckers of around 100 guns, the First Rates, directly descended from the *Sovereign*; slightly smaller three-deckers of 90 guns of similar origin; and two-deckers of 70 guns, produced by a gradual expansion of the *Speaker/Mary* design.

The British Fall Behind

Though the British played the major role in the initial development of the ship of the line they were soon to lose their lead and by the middle of the eighteenth century they had fallen well behind. In the 1690s they concentrated on other types of ship – small three-deckers of 80 guns, and two-deckers of 50 or 60 guns. Later experience was to show that the best ships had a large number of guns for their number of decks – 100-gun three-deckers were better than 90s or 80s, and 70-gun two-deckers were better than 60s or 50s. A small two-decker, or a small three-decker, tended to be ill-proportioned and too high out of the water in relation to her length,

the Roy.ll Charles; And the Dutch Fleet commanded by Adm.ll de Ruyter on the 25th July 1666 Together wth a List of the English Shipps & Capt wth their numbers of Men & Guns

The line of battle in the St James's Day Fight, 1666, in the Second Anglo-Dutch War. (National Maritime Museum)

The Sovereign of the Seas *as built in 1637. The first true three-decker in the world. (The Science Museum)*

making her a poor sailer and essentially unstable. Furthermore, it was eventually proved that a small number of large guns was better than a large number of small ones; the optimum gun was the 32-pounder, able to penetrate the enemy's sides far better than lesser guns. Heavier guns, such as 42-pounders, had balls that were too big to handle in action.

The British navy was large and undefeated as it entered the eighteenth century but its ships were generally poorly designed. The three-deckers of 90 and 100 guns were powerful but clumsy to sail, expensive to build and difficult to man. The three-decker 80s were the worst ships in the fleet, being unstable and poor sailers. The two-decker 70s were probably the best ships but they were relatively lightly gunned, with only 24-pounders on the lower deck. The 50s and 60s were even weaker, having only 18- and 12-pounders as their main armament. For the moment this did not matter much as the other large navies, the French, Dutch and Spanish, were mostly inactive.

In the 1720s and 30s both the French and Spanish navies began to revive and introduce some fresh ideas on ship design. The French navy no longer expected to be able to match the British ship for ship at sea so instead concentrated on making each individual ship as effective as possible. In these circumstances it evolved the 74-gun ship, the type that was to dominate the battlefleets at least until the second quarter of the next century.

British ship design on the other hand, had by 1720 become rigidly conservative. Resistance to change permeated the whole system and it would be difficult to single out any individual who was responsible for this state of affairs. Nevertheless, Sir Jacob Ackworth, Surveyor of the Navy from 1715 until his death in 1749, was to attract much of the criticism of those who wanted to reform the navy; according to Admiral Vernon he was 'a half-experienced and half-judicious surveyor' who had 'half-ruined the navy'.[2]

In the first place, the sizes of ships were controlled by the 'establishment of dimensions'. This system had begun rather modestly in 1705–6 when the Admiralty had simply taken the principal dimensions of the most successful ship of each type and ordered that subsequent ships should be built to them. The system was reinforced in 1719 by means of a new establishment. While this made only slight changes to the main dimensions it also established all the dimensions and scantlings of future ships, down to the smallest vessels – the sizes of hatches, the width of individual timbers and beams, and the thicknesses of the planks – which were to be used. Only the actual shape of the hull, the lines, was still left to the ship designer. The establishment was not immutable but changes depended on a certain measure of agreement between the Admiralty and the Navy Board, the latter of which was responsible for the more technical matters. Intelligence of the improved French and Spanish ships caused a new establishment to be proposed in 1733 but this was never fully ratified though many ships were built to it.

Second, British warships were now 'rebuilt'. In theory at least, old ships were not scrapped and replaced by new ones but kept on the navy list, taken to pieces and any serviceable timbers incorporated in the new vessel. The new ship usually had the same name as the old one and the same number of guns; but the amount of old timber incorporated in a rebuild declined greatly over the years. The system of rebuilding allowed some updating of design and rebuilt ships always followed the current establishment, but it discouraged any radical developments.

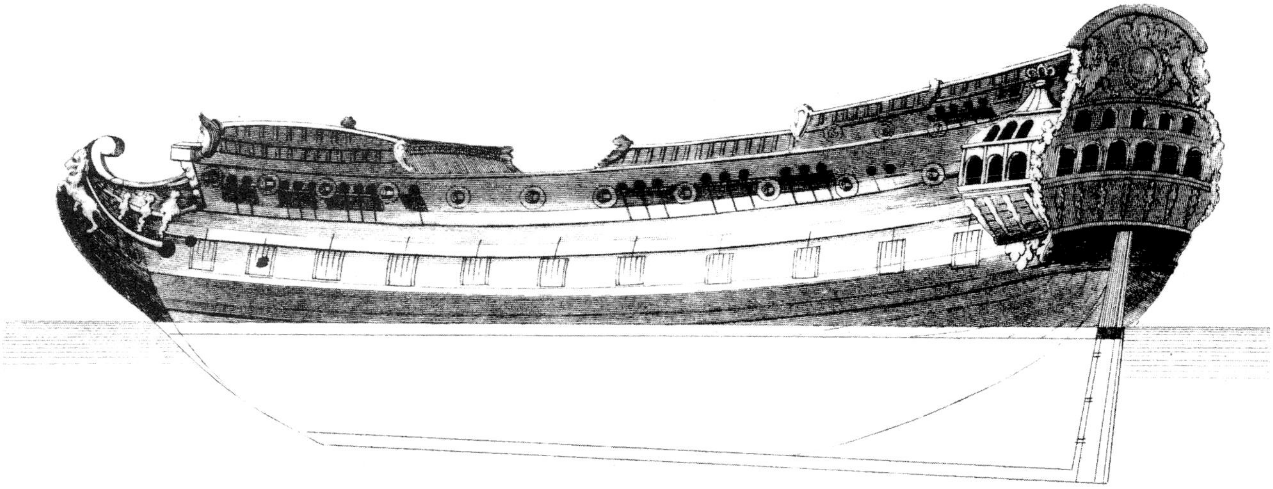

The Speaker *of 1649, the prototype of the British two-decker. (From Charnock's* History of Marine Architecture*)*

Types of ship of the line in the British fleet, c.*1730*

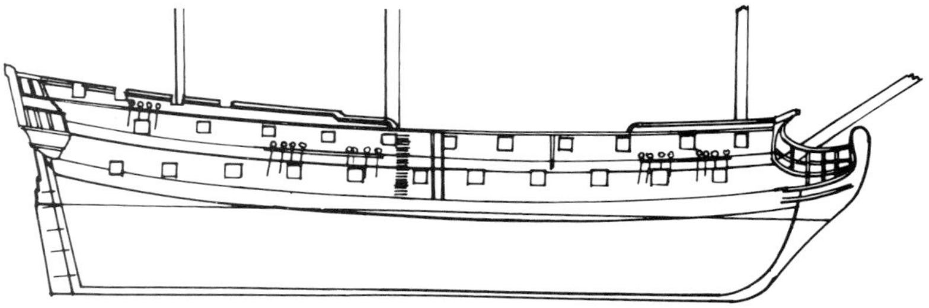

50-gun two-decker

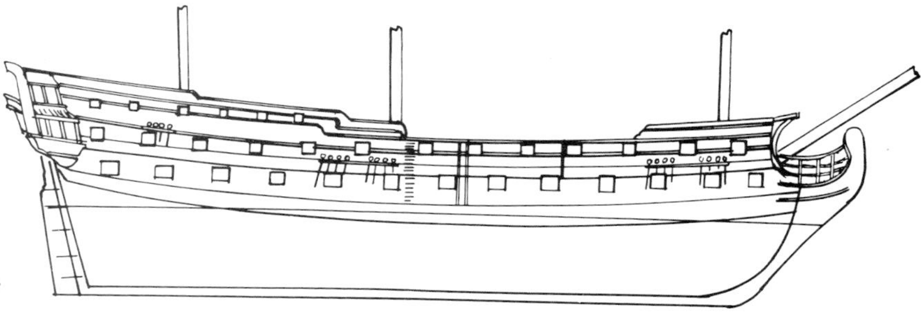

60-gun two-decker

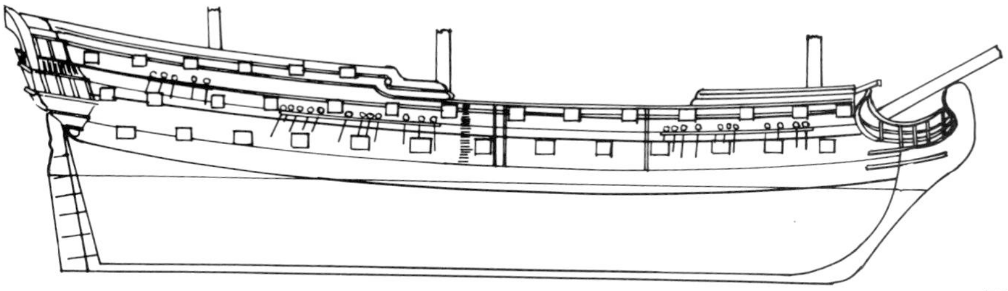

70-gun two-decker

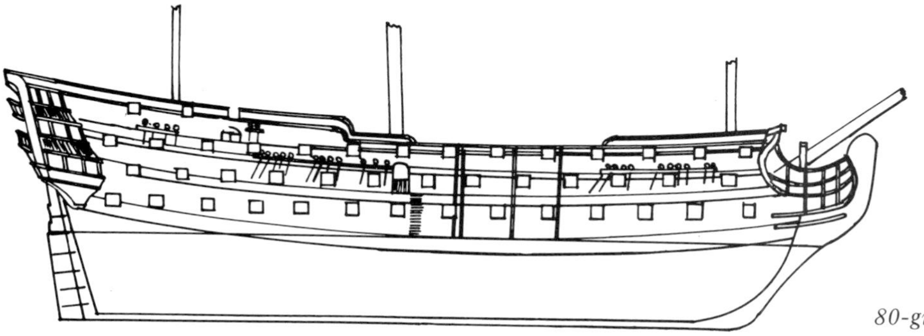

80-gun three-decker

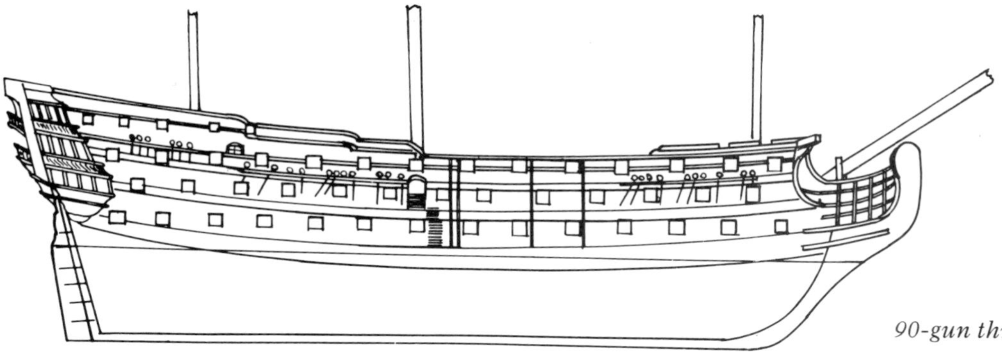

90-gun three-decker

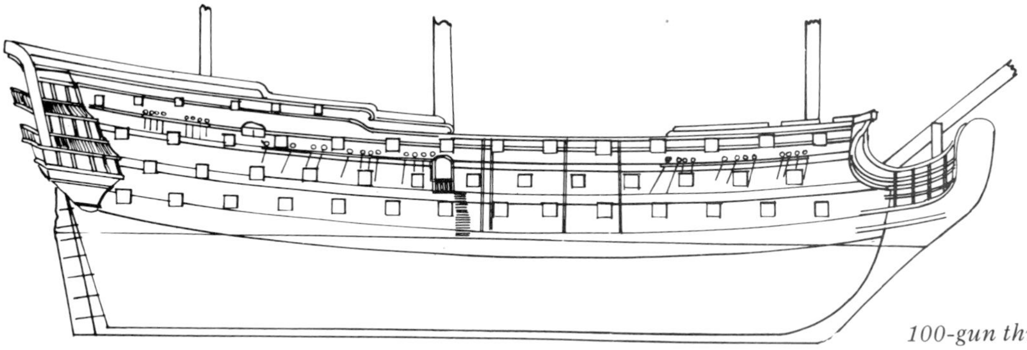

100-gun three-decker

George Anson.
(National Maritime Museum)

A model of an 80-gun ship of the 1719 Establishment showing the short hull and high sides which made such ships unstable and poor sailers. ➤
(The Science Museum)

Last, naval administrators decreed that any new ship, whether a rebuild or a replacement for a ship lost at sea, should be of the same form and have the same gun power as the old one. Thus the composition of the fleet changed very little between 1714 and 1739; except that a few of the 50-gun ships were replaced by 60s. Design was almost frozen, and unsuccessful types, such as the 80-gun three-deckers, were kept in service despite their defects.

The Faults of British Ships

War broke out with Spain in 1739 and with France in 1744. The Spanish navy was small and no battlefleet was fitted out; nevertheless, many of the individual Spanish ships were excellent and in 1740 a 70-gun ship, the *Princessa*, put up a long fight before being captured by three British ships of nominally equal force. It was soon pointed out that the *Princessa* was equal in size to a British 90-gun three-decker and British ship designers were forced to rethink; another establishment was proposed which included increases in dimensions. In addition, some classes were to be given heavier types of gun and in slightly reduced numbers. Thus the two-decker 70, with 24-pounders on the lower deck, was replaced by the two-decker 64, with 32-pounders. This was a step in the right direction but did not go far enough for nothing was done to replace the most unsuccessful classes such as the three-decker 80s. On the other hand, the system of rebuilding soon collapsed under the pressure of war for dry docks were needed for hasty repairs rather than long rebuilds.

Meanwhile, complaints about the quality of British ships continued to pour into the Admiralty, from admirals and captains. Of the three-deckers, Admiral Mathews complained, 'I have now but two ships of 90 guns and three of 80 guns that can make use of their lower tier of guns if it blow a cap full of wind . . . As for the rest of them, they can scarce haul up a port'. Admiral Vernon pointed out that the large three-deckers were useless for any sort of coastal defence as they needed too many men and could not be used in shallow waters. Of the other ships, one officer noted, 'the general discharge of an English 80-gun ship is 1,312 pounds; a French 74 is 1,705 pounds; an English 70 is 1,044 pounds; an English 60 is 918 pounds; a French 64 is 1,103 pounds'. Commodore Knowles observed that 'it is pretty apparent that our 70-gun ships are little superior to their ships of 52 guns'.[3]

Anson and the 1745 Establishment

Admiral George Anson first took office as a member of the Board of Admiralty in 1744. He was already a national hero and a rich man having led a four-year circumnavigation against the Spanish colonies, captured a treasure galleon and surmounted unimaginable difficulties with both men and ships to reach ultimate success. Anson knew the faults of British ships as much as anyone else, and was determined to do something about them; but he suffered fierce resistance from Ackworth and the officials of the Navy Board and from some of his fellow sea officers. In 1744 yet another new establishment was proposed, but this time every effort was made to produce a more radical conclusion than in the past. A committee of flag officers and senior captains was set up to consider proposals by the leading shipbuilders. The troubles of recent years were to be taken into account. 'It is also a general complaint that the ships are crank, and heel so much in blowing weather that they cannot open their lee ports'. The 80-gun ship was to be abandoned; 'as the present ships of 80 guns, with three decks, are in general ill approved, they were directed to propose ships carrying 74 guns, with two decks and a half in their room'.

In fact, the report, produced in 1745, was a disappointment. The sizes of ships were to be increased

The capture of the Princessa *in 1740.*
(National Maritime Museum)

but not by nearly enough to match foreign construction; the committee had inquired about the depth of water in the dockyards and concluded that ships could not be made much bigger. The committee rejected proposals for 74-gun ships, being 'sorry to differ with your lordships therein, but we having observed on many occasions the advantage which 80-gun ships with three decks had over those with two and a half, judged it for the benefit of the service that so useful a class of ships should be continued'.[4] The final upshot was that the system of establishments became more rigid than ever. Even the lines of the ships were now fixed, leaving virtually no initiative to future designers; and the establishment was ratified by the Privy Council so that the control of ships design was no longer even within the naval administration.

However, the Admiralty did not intend to give up the 74-gun ship entirely. A few were built, mostly by converting 80s and 90s of the old establishments. More important, several were captured from the French during the battles of 1747, notably the *Invincible, Terrible* and *Monarch.* Though mainstream British ship design was still unable to deviate from the principles of the establishment, the Admiralty had no intention of keeping up the old 80s. The type remained on the establishment but there was no compulsion to build any and only two were begun after 1745.

The war with France and Spain ended in 1749. Ackworth died in the same year but his successor, Sir Joseph Allin, was little more imaginative. The ships of the 1745 establishment proved to be unsuccessful and captains complained that 'they do not steer so easy, nor sail so well, as was expected'. Small changes in design were proposed, and each year the Admiralty had to take its plans to the Privy Council to have every variation from the 1745 establishment approved. Anson rose higher in the naval administration, and became First Lord of the Admiralty in 1751, but he was unable to do anything further to advance ship design.

The New 74s

The breakthrough came in 1755, on the eve of yet another war with France. Sir Joseph Allin suddenly became ill and 'disordered in his senses'. The Admiralty immediately ordered Thomas Slade and William Bately to take his place as Joint Surveyors of the Navy. Within three weeks of his appointment Slade had produced a new design for a two-decker, somewhat larger than those of 1745. For a time the Admiralty persisted in calling it a 70-gun ship though a count of its gunports would easily reveal that it was intended to carry 74. The fiction was kept up for several months before the Admiralty was prepared to describe the ship as a 74. Perhaps it feared that such a radical breach of the priniciples of the 1745 establishment would reduce the government's confidence in them; perhaps it was felt that the 74 represented everything the old surveyors and administrators had been against, for it was a large two-decker of essentially foreign origin. Both Anson and Slade were extremely taciturn at the best of times and committed little to paper, so probably we will never know.

The new 74s, of the *Dublin* class, were the first such ships to be built from the keel up in Britain. Yet they were only a limited advance on what had gone before. They were 5½ft longer than the standard 70s of the 1745 establishment, and 3½ft longer than the largest ships built to variations of that establishment. They were measured at 1,560 tons, compared with 1,414 tons for the 1745 establishment, and around 1,000 tons for a large French 74 like the *Invincible* or *Magnanime.* Over the next few years Anson and Slade began to gain in confidence and expand the 74s. The *Hero, Hercules* and *Thunderer* were ordered in 1756 and were 1ft longer than the *Dublin* class. In the following year three ships of the *Bellona* class were ordered measuring 168ft long and of over 1,600 tons. These ships were to set the pattern for British 74s over the next twenty years and though they were still 200 tons less than the largest French ships they were substantially larger than the ships of the 1745 establishment, not to mention those which had gone before. It was only with the building of the *Valiant* and her sister ship *Triumph*, also ordered in 1757, that British-built 74s measured up to those of the French; but by that time the 74 was established as the standard ship of the line in both the British and the French navies.

What made the 74 so successful? In essence it was the ideal compromise between all the factors which needed to be taken into account in the design of a ship. She was the smallest successful ship to carry a full battery of 32-pounders, and that was the most successful type of naval gun. She was a two-decker, so her centre of gravity was lower than on a three-decker, and therefore she was more stable. Furthermore, her sides were lower, and this was a great advantage when sailing into the wind or with the wind on the beam. According to one leading naval architect, 'The 74-gun ship . . . contains the properties of the First Rate and the frigate. She will not shrink from an encounter with a First Rate ship on account of superior weight, nor abandon the chase of a frigate on account of swiftness. The union of these qualities has therefore, with justice, made the 74-gun ship the principle object of maritime attention, and given her so distinguished and pre-eminent a place in our line of battle.'[5]

2

The *Invincible* and the *Valiant*

While the British navy was stagnating in the 1730s and 40s the French entered its most creative phase. Forty years earlier, under Louis XIV, France had built up a fleet to challenge for the supremacy of the seas. That force had been defeated at Barfleur and La Hogue in 1692, and the French had come to terms with the fact that they were not really a sea power in the same sense as the British; that they had a vulnerable land frontier to their north east; that they had fewer good ports and a smaller seafaring population, and that their economic strength was not sufficient to support both a large army and a strong navy.

The first step had been to privatise naval warfare and rely on privateers to raid enemy commerce. This was of considerable nuisance value but at that period Britain was not a major importer of either food or raw materials and she could not ultimately be defeated this way. A further step was to cut back the navy and as a result the fleet was gradually reduced in numbers after 1692. In the war of 1702–14 the French battlefleet made no serious challenge at sea and by 1720 the navy had declined to a total strength of 40 vessels of all types, at a time when the British had about 240.[1]

The French Revival

By 1730, however, the French navy was beginning to find a new role for itself. The major wars of the mid-eighteenth century were to be fought largely over colonies and such warfare demanded ships which could sail anywhere and stay at sea for long periods. Previous naval wars had largely been fought in European waters between ships which were fitted out for the summer only and rarely went far from their home bases. Future wars would be fought in all the oceans of the world, for possession of North and South America, the Caribbean Islands, India and bases round the world. In the seventeenth century the colonies, for example North America, had been useful as dumping grounds for adventurers, criminals and religious fanatics: by the mid-eighteenth century, they were vital and expanding sectors of the European economies. A fleet of fast, powerful but seaworthy ships could threaten to appear anywhere; to launch an invasion of a West Indian Island or reinforce an army in India or Canada. It could even threaten an invasion of the British homeland itself, perhaps in alliance with Scottish or Irish rebels. With good ships and sound strategy it did not need total superiority over its enemies; with local superiority it could achieve much, and it could act as a permanent threat to the British position in the world.

The Origins of the 74

The French had no preconceived plan to develop the 74 as the main ship of the line. Instead, the Minister of the Marine, Maurepas, in absolute contrast to the situation across the Channel, simply gave the shipbuilders a free hand. Three-deckers were almost universally rejected, but many types of two-decker – of 80, 74, 72, 70, 66, 64, 60, 56 and 50 guns – were begun in the 1730s and 40s. Up until that time an old type of 74 had been built, similar to the British 70 except that it had four very light guns on the poop. In 1737 François Coullomb, the master shipwright at Toulon, began a new type of 74. It had 36-pounders on the lower deck instead of 24-pounders and its guns on the poop were abandoned in favour of much heavier guns on the main decks. To support the greater weight of armament it was lengthened to 156ft (French) which was considerably longer than the older type of 74. This ship, the *Terrible*, was reasonably successful, though hindsight suggests that the increase in length was not sufficient.

Two more 74s of the new type were laid down at Rochefort in 1741. One, the *Magnanime*, was to be built by Geslain, the chief constructor, and the other, the *Invincible*, by his assistant, Pierre Morineau. It is often said that the French achieved superiority in shipbuilding because of their more scientific training, but there is no real evidence for this. Although amateur scientists paid more attention to these matters in France than in Britain, it did not filter down to shipbuilders in any significant or practical way. Certainly the French were the first, by seventy years, to set up a School of Naval Architecture; but this was founded in 1741, the same year as the *Invincible* was begun, and clearly had no effect on the design of that

Types of ship of the line in the French navy, 1720-45.

80-gun two-decker

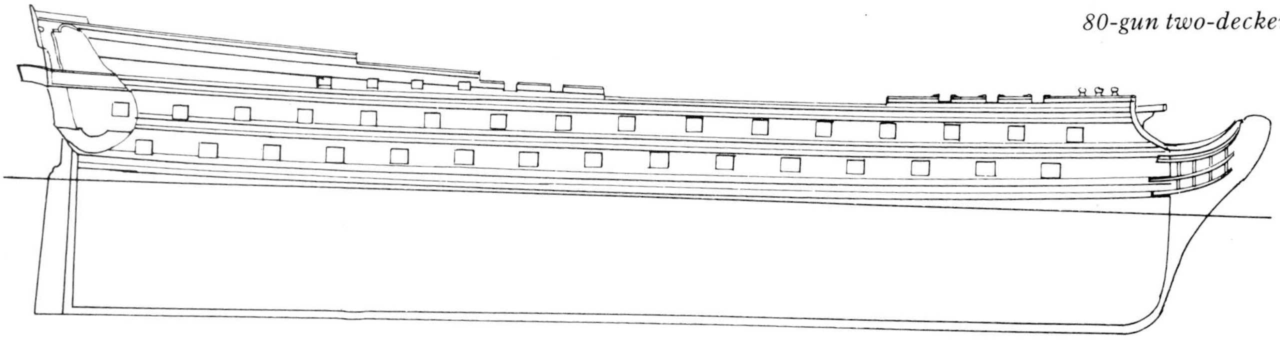

74-gun two-decker, similar to the Invincible

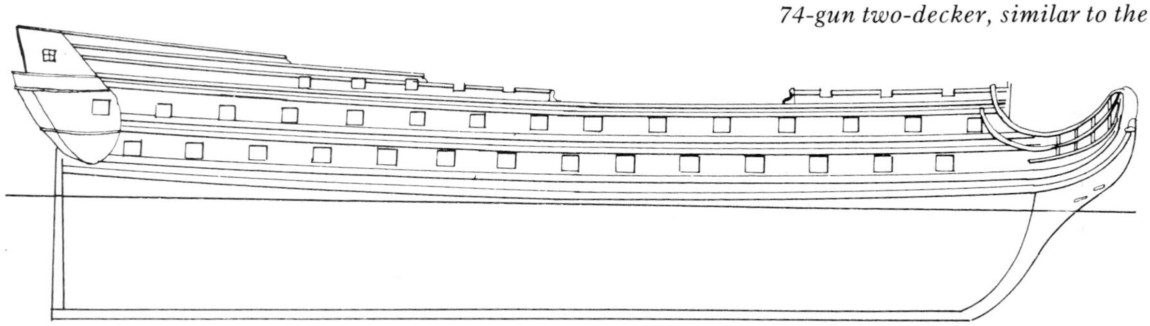

The old type of 74-gun ship, with four guns on the poop

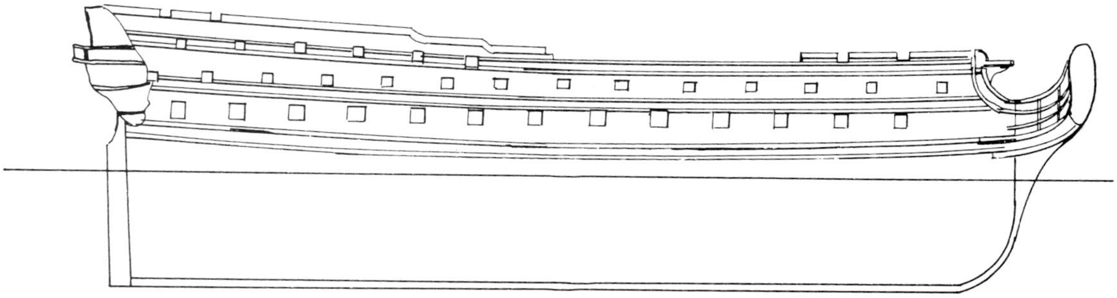

70-gun two-decker, as superseded by the 74

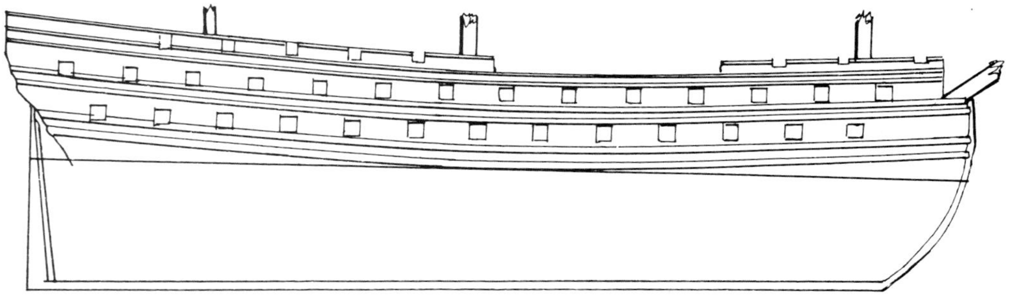

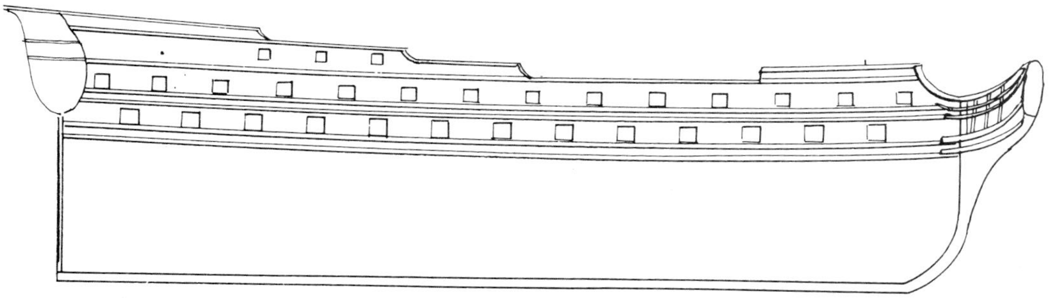

64-gun two-decker

ship. Not much is known about Pierre Morineau beyond the posts he held, but he almost certainly served an apprenticeship in the traditional manner, to a master craftsman in one of the royal dockyards. On the other hand, he wrote two manuscript treatises on ship design, which suggests that he was fully literate, and that he was rather less secretive about his methods than most eighteenth-century craftsmen.[2]

The plans of the *Invincible* were sent to Paris for approval in September 1741 and construction began around the end of October. The new ship was to be 162ft (French, 171ft English), which was 6ft longer than the *Terrible* and nearly 20ft longer than a typical British 70 of the time. She was expensive to build – in December 1743 Morineau complained that she was expected to cost 150,000 *livres* compared with 70,000 for the *Terrible* – but events were to show that she was a particularly well-designed ship, and the cost was justified.

The *Invincible* in the French Navy

The *Invincible* was launched at Rochefort on the 21 October 1744 and fitted out in the River Charente and in the anchorage of Ile d'Aix. Her seventy-four guns of brass and iron, and her masts and rigging were installed, and she was ready for sea in July 1745.

Her first service was to escort a convoy to the French West Indies. While in the Caribbean, sailing between Haiti and Cuba, she had her first encounter with the British. Shots were exchanged and the *Invincible* suffered some damage to her rigging, though the British captains were impressed with the size of her guns. She returned to Rochefort in August 1746 and was refitted in the dockyard there.

It was decided that the *Invincible*, along with two other ships, should be leased to the French East India Company to escort an important convoy to India. She put to sea from L'Orient in March 1747 under the command of Captain St Georges and soon had another encounter with the British. St Georges drew up the largest ships of his convoy in a line of battle which convinced the British commander, Captain Savage Mostyn, that the fleet was strong for through the mist Mostyn had been convinced that he had seen a three-decker and eighteen smaller warships. Having eluded his enemies St Georges had less luck with the weather and he was eventually forced back to Rochefort due to storms.

There, St Georges's India convoy met with another one, bound for Canada, under Admiral La Jonquiere. The two were combined, and set sail early in May. They were soon sighted by a strong British force under Admiral Anson. La Jonquiere tried the same trick as St Georges – he drew up his largest ships in line of battle and ordered the smaller ones to escape. Anson was a shrewder commander than Mostyn though, and had, in addition, a much larger fleet. He was not fooled by the French tactic and the First Battle of Finsterre ensued.

As the British fleet approached the captains of the merchant ships lost their nerve and ran while the rest of the French squadron were attacked one by one as the British caught up with them. The *Invincible* was one of the last to be engaged and she held her own for some time. But she suffered heavy casualties – 80 men were killed and 170 wounded out of a crew of 700 – and she was eventually surrounded by three British ships and ran out of ammunition. It is said that St Georges even ordered the cutlery from his cabin to be fired at the enemy but she was forced, nonetheless, to surrender and a prize crew from the *Pembroke* took possession of her. St Georges was taken aboard Anson's flagship, the *Prince George*, and the two men soon became friends.

Anson was immediately impressed with the *Invincible*. He sent his carpenter aboard to measure her, and commented that she was 'a prodigious fine ship, and vastly large'. His regard for the ship was to be maintained over the years, and he seems to have taken a great deal of interest in her. The ship had lost much of her top hamper in the battle and had to be towed back to Portsmouth. She was bought for the British Navy for £23,738 which was divided among the captors of the ship, with the lion's share going to the senior officers. The battle

L'INVINCIBLE

The bow and stern of the Invincible. *(From Charnock's* History of Marine Architecture*)*

Rochefort dockyard in the eighteenth century, as painted by Vernet. (Musée de la Marine, Paris)

The First Battle of Finisterre. (National Maritime Museum)

damage was repaired and she was given new masts and rigging, guns, cabins, stores and other fittings, and made ready for service in the British navy.

The *Invincible* in the British Navy

The *Invincible* served for eleven years in the British fleet, seeing a good deal of service but very little real action with the enemy. In 1748 she was with the Channel Fleet as flagship for part of the time. The war ended that year, and she was made a guardship at Portsmouth where she stayed until being fitted out for a voyage to Gibraltar in 1752. After that she was found to be in poor condition and her hull was taken apart for a major repair. The Board of Admiralty issued special instructions to repair the ship rather than scrap her though the cost of the repair was at least as great as that of a new ship. In

Admiral Edward Boscawen.
(National Maritime Museum)

1757, during the early years of the Seven Years War, she put to sea again and served for a time as flagship of the Channel Fleet under Admiral Boscawen. In the following year she crossed the Atlantic as part of a major expedition against the French stronghold at Louisbourg and like most of the ships in that fleet was caught in a hurricane and severly damaged. She returned home for repairs and in February 1758 she set sail, as part of a fleet under Boscawen, for another attack on Louisbourg. While beating out of the Solent she ran aground on the Dean Sand and was unable to get herself off. Stores were taken off her, but attempts to float her were unsuccessful and had to be abandoned after three days. She slowly broke up, but in 1979 her wreck was rediscovered by Arthur Mack, a local fisherman. Many objects have been recovered from her, and the best of these are now on display in the Historic Dockyard at Chatham.

The Influence of the *Invincible*

During the period of her service with the British fleet, the *Invincible* had considerable influence on ship design. French prizes were much less common than they were to become fifty years later, in the days of Nelson. British ship design was in a particularly bad way and the contrast with French design as represented in ships like the *Invincible* was striking. To Admiral Warren, who was present at her capture and used her as his flagship in 1748, the *Invincible* was 'better in every way than any ship, and is in every shape a fine man of war'. To Captain Keppel, 'the *Invincible* outsails the whole navy of England'. To Boscawen, 'the *Invincible* sails well, rather better than every ship'. Captain Bentley, who commanded her on the voyage to Gibraltar and during her loss, reported that she could make 13 knots in ideal conditions, whereas few British ships of the period recorded speeds of more than 11 knots.

It is not surprising, therefore, that the idea of copying the *Invincible* occurred soon after her capture. In August 1747, shortly after the ship had been surveyed at Portsmouth, the Admiralty attempted to bypass the 1745 establishment. It pointed out that 'In the late establishment for building a ship of each class in the Royal Navy' there was 'no mention of a ship of two and a half decks to carry 74 guns', and the Navy Board was ordered to 'consider of proper dimensions for such a ship to carry that number of guns, and when they have done, to propose a draught and solid [i.e. model] by which they propose to build her'. The draught and model were ready by January 1748.[3] Neither has survived so it is not known how similar they were to the *Invincible* but surviving information suggests that the design was close. It allowed for fourteen ports per side on the lower deck and fifteen on the upper, and this was a traditional French feature, compared with fourteen ports on both decks of a British ship. The Admiralty ordered that two ships be built to the design, one at Woolwich and the other at Chatham, but both were cancelled for lack of funds when the war ended a few months later.

No more was to be heard about copying French design for some time until money was made available for shipbuilding in 1755 as another war with France threatened. However, Anson now had his own men, Slade and Bately, at the Navy Board and they began to develop a British style of 74-gun ship, different from the French style as represented by the *Invincible*, but clearly much influenced by it. As the *Dublin, Hero* and *Bellona* classes were laid down, the length of the British 74 began to approach that of the *Invincible*, the *Bellona* class being 168ft on the gundeck, compared with 171ft on the *Invincible*.

The Invincible *after her capture.*

The Order for the *Valiant*

The idea of copying French design was revived in May 1757. Anson was not on the Board of Admiralty at this stage as he had fallen temporarily from office because of the scandal over the loss of Minorca and the execution of Admiral Byng. The most influential naval member of the Board was now Admiral Boscawen. He was a great friend to the *Invincible* for he had spent three happy months aboard her in the previous year when she had served as flagship of the Channel Fleet. He had written to his wife, 'I should have told you, the *Invincible* sails well, rather better than every ship, nothing but the *Bedford* comes near us'. Even her accommodation was praised, and her admiral's cabin was 'strong and plain like an old country hall'. When he shifted his flag to the First Rate *Royal George*, he found that 'she is not so convenient in many particulars as the *Invincible*, though fitted for Lord Anson'.[4]

There can be little doubt that Boscawen had much influence on a decision taken by the Admiralty Board on 21 May 1757. Two ships had been ordered in January, to be built at Chatham and Woolwich, to the same draught as the *Dublin*. It was observed that work had not yet been started on them, so the plans were changed. 'Experience having shown that His Majesty's ship the *Invincible* is in every respect the best ship of her class, and answers all purposes that can be desired in a ship of war, we do hereby desire and direct you to cause the two beforementioned ships to be built by the draught of the *Invincible*, and in every respect of the same scantlings, notwithstanding any former orders to the contrary.'[5] The new ships had already been named – the one at Woolwich was to be called the *Triumph* and the Chatham ship was to be named the *Valiant*.

3

Drawing the Lines

For nearly two centuries all large ships, especially those for the navy, had been built from plans drawn on paper. In the more distant past all shipbuilders had worked by 'hand and eye', and had produced their ships without any plans in the modern sense. This method had become impracticable as ships became larger, though builders of smaller vessels continued to do without plans even into the twentieth century. The earliest known English scale plans are in a volume entitled *Fragments of Early English Shipwrightry*, dated 1586, in Magdelene College, Cambridge. By about 1715 it was common to send copies of all plans for naval ships to the Navy Board in London for approval; most of these were retained, and now form the basis of the great collection of plans at the National Maritime Museum, Greenwich.

The Plans

For the first half of the eighteenth century plans, or draughts as they were usually called at the time, were drawn by the Master Shipwrights in charge of the various yards. From 1745–55 the plans were done exactly in accordance with the Establishment of Dimensions, and after that most plans originated from the office of the Surveyor of the Navy in London. Three or four draughtsmen were employed in the Navy Office for this work, though one or other of the two surveyors exercised a close supervision over each design, and signed each plan when it was complete.

A ship's draught was drawn to a scale of 4ft to 1in or 1/48. It included four separate views of the ship. The profile, or sheer draught, showed the ship from the side, with the keel, stem and sternpost, gunports, channels, wales and an outline of the decoration on the bow and stern, though without any detail of the figurehead and other carvings. The heights of various decks might be indicated, or the detail of the internal construction might be drawn in red ink, contrasting with the black of the rest of the draught.

The second and third parts, drawn side by side, were the body sections of the hull. Two views were taken, one from the bow and one from the stern. Cross sections of the hull, usually at every third frame or rib, were drawn, to give the first indication of the shape of the lines of the ship. The fourth part of the draught, drawn below the sheer draught, was the half breadth plan, which showed the line of maximum breadth, along with water lines, buttock lines and other features. These were used to 'fair' the design – to check that the curves produced by the various shapes of the frames would be reasonably smooth when the ship was finished.

Other plans might be added when necessary. If the internal details were not included in red on the sheer draught, then a separate longitudinal section might be drawn showing the positions of decks and their supporting beams, capstans, the cooking stove, hatchways and ladders, pumps, masts and some of the internal compartments such as the magazines and store rooms. Deck plans were quite common by this time, and in most cases a single sheet showed two or three decks: the orlop and lower deck, or the upper deck combined with the quarterdeck and forecastle. Positions of deck beams would be marked, along with hatchways and bulkheads.

When completed, the plan would be copied as necessary. There was no easy means of doing this but usually the draughtsman pricked through the original with a pin so that key points could be reproduced on another sheet underneath. The master copy was kept in the Navy Office and a copy was sent to each yard which was building a ship to this design – the Chatham and Woolwich in the case of the *Valiant* and *Triumph*. It was usually sent by coach, having been placed inside a specially designed box, and the draught of the *Valiant* was sent to Chatham on 21 July, exactly two months after the Admiralty ordered the ships to be built to the draught of the *Invincible*.

Because the *Triumph* and *Valiant* were copied from a French ship, the Surveyor of the Navy and his draughtsmen had rather less freedom than usual. The standard practice for designing ships is described here, though it must have differed somewhat in the case of these particular vessels.

Several important features of the design had already been fixed by Admiralty Order. The length of the ship on the gundeck was laid down, as was the maximum

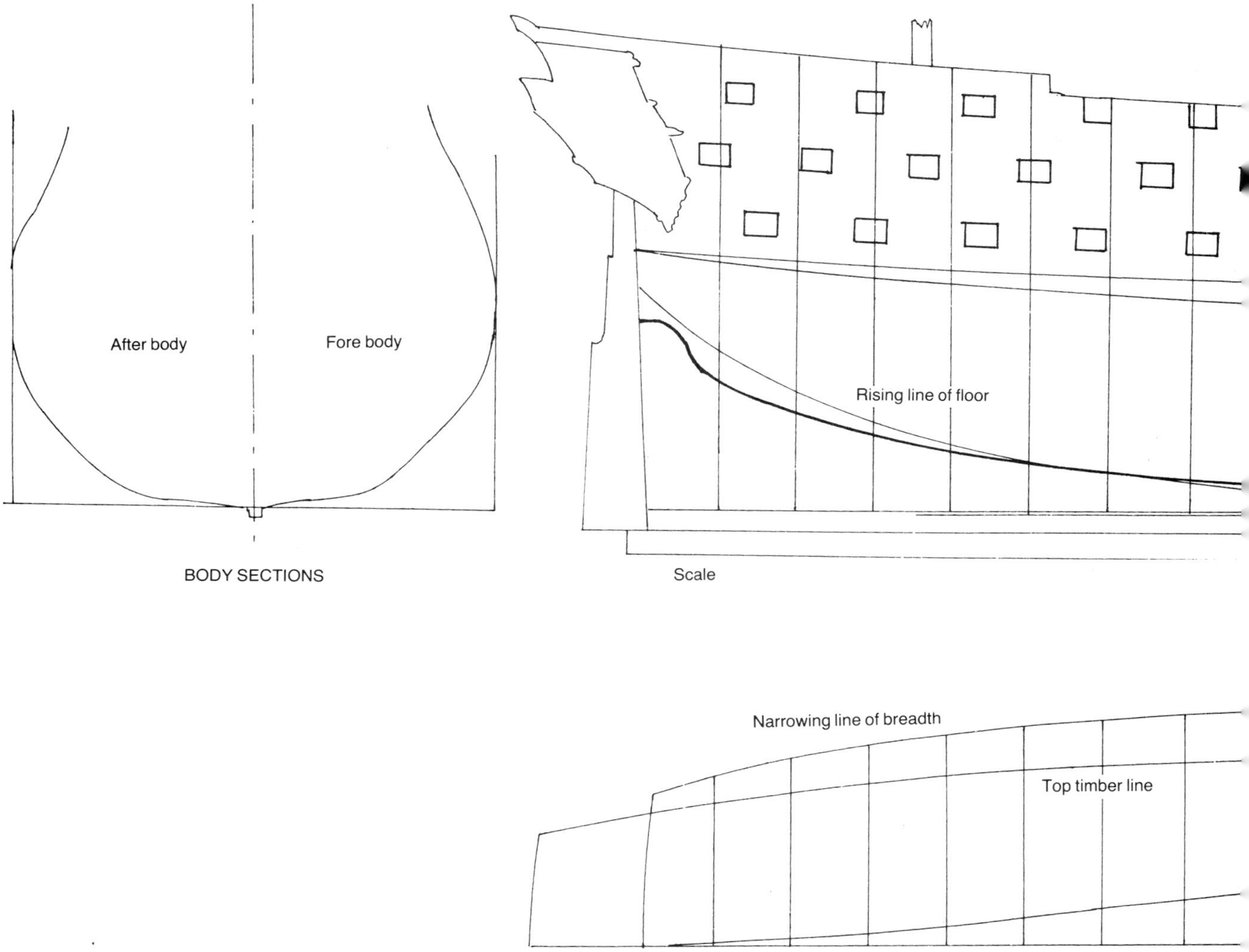
After body
Fore body
BODY SECTIONS
Rising line of floor
Scale
Narrowing line of breadth
Top timber line

SHEER DRAUGHT

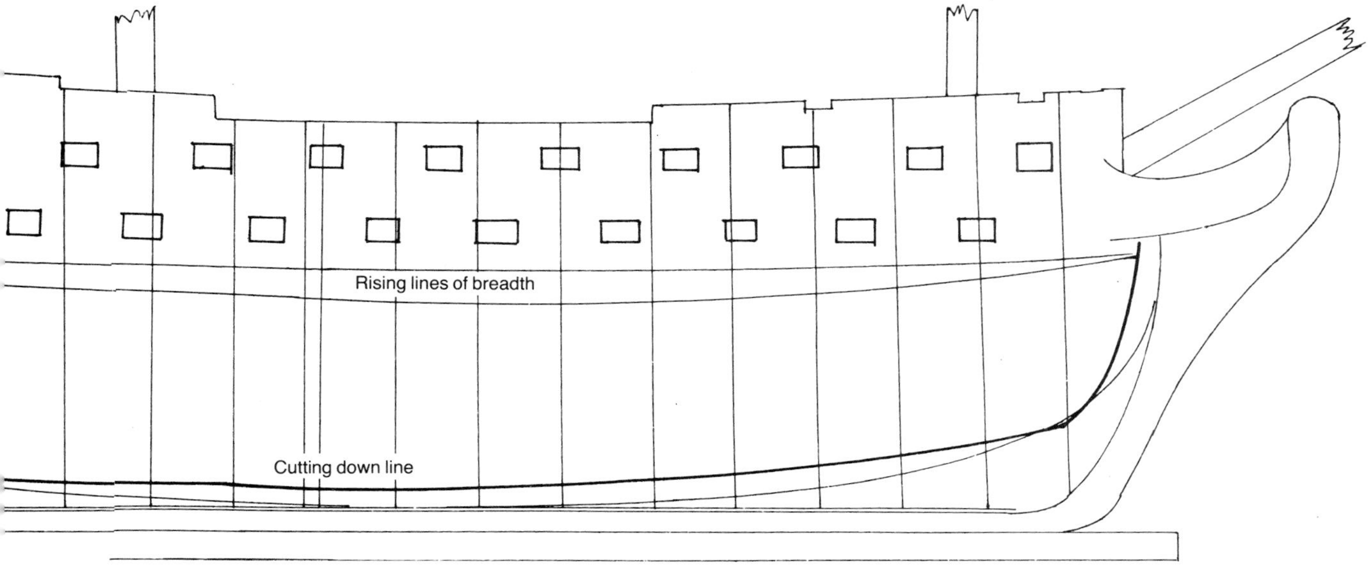

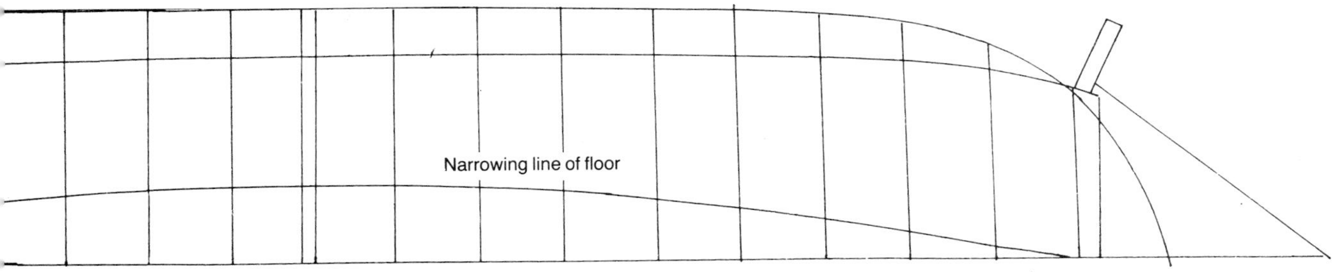

HALF BREADTH PLAN

The draught of the Valiant, *showing the sheer draught, half breadth plan and body sections.*

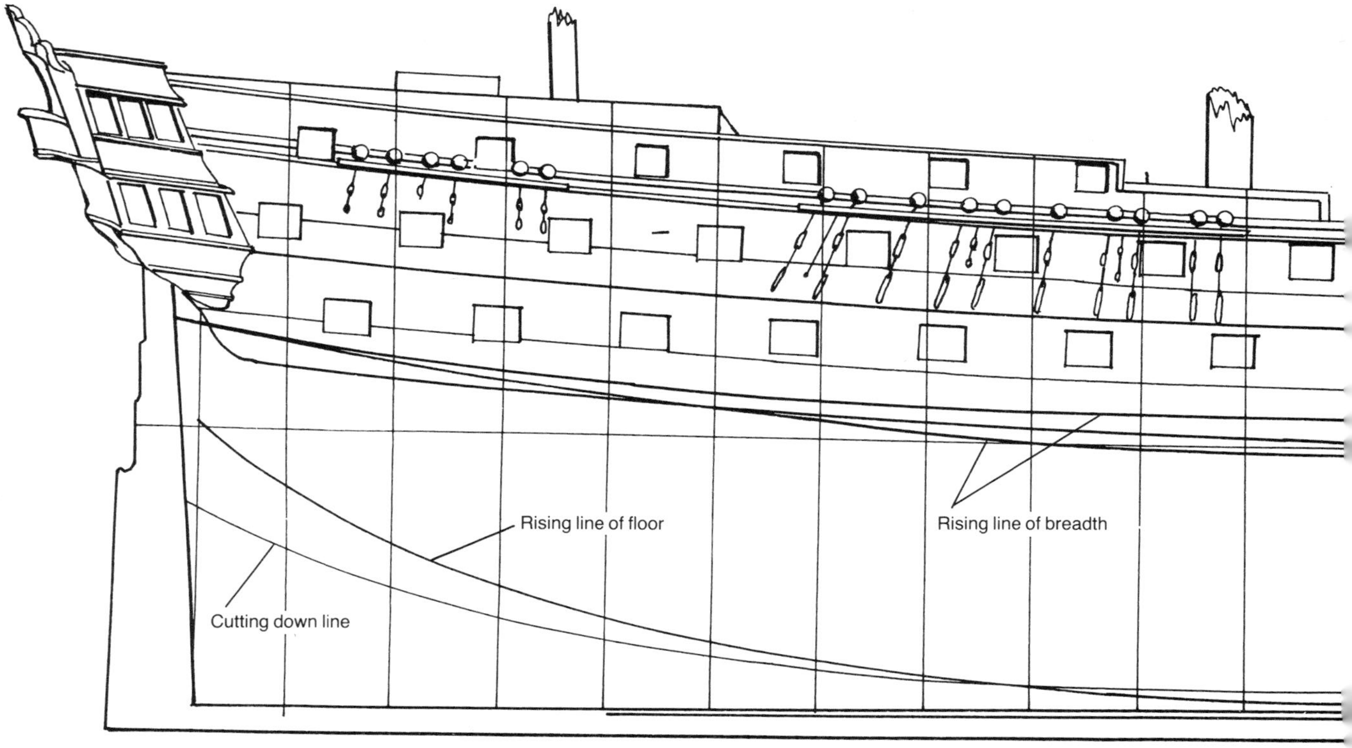

Details of the sheer draught, with the rising and cutting down lines.

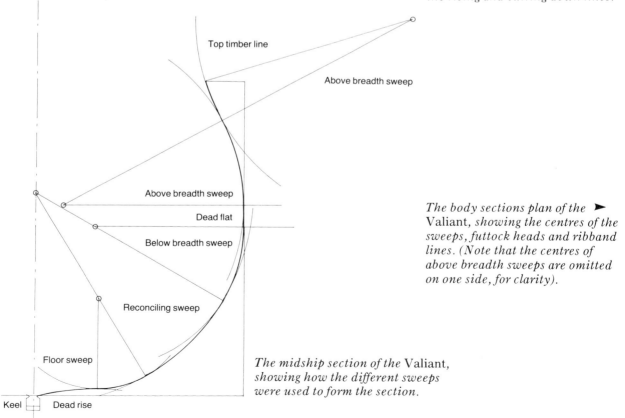

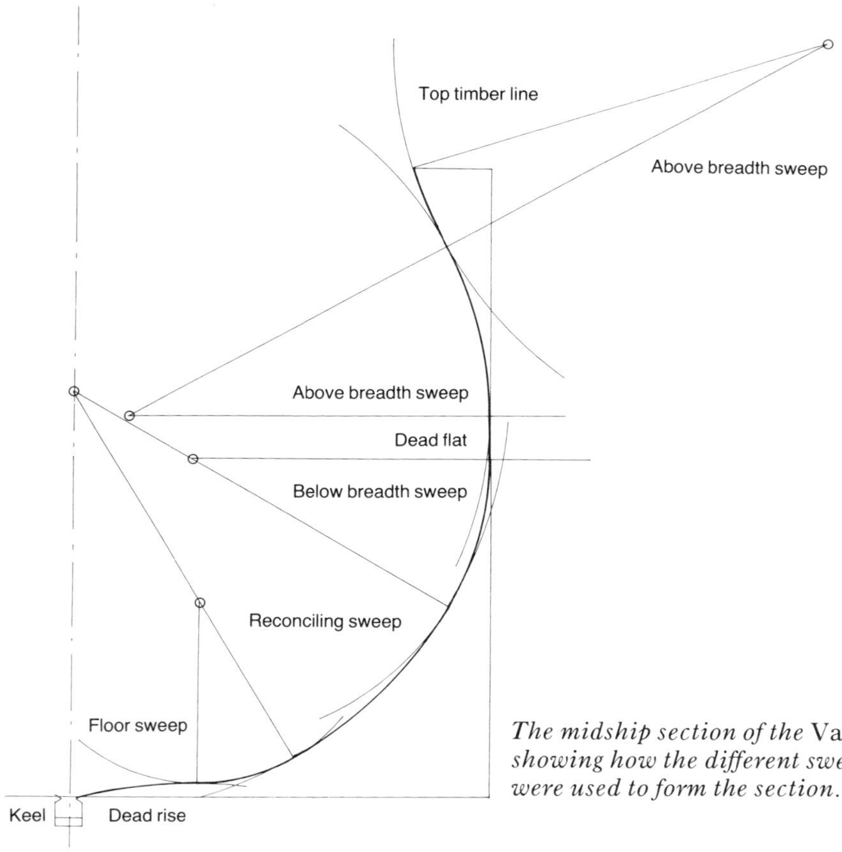

The body sections plan of the ➤ Valiant, *showing the centres of the sweeps, futtock heads and ribband lines. (Note that the centres of above breadth sweeps are omitted on one side, for clarity).*

The midship section of the Valiant, *showing how the different sweeps were used to form the section.*

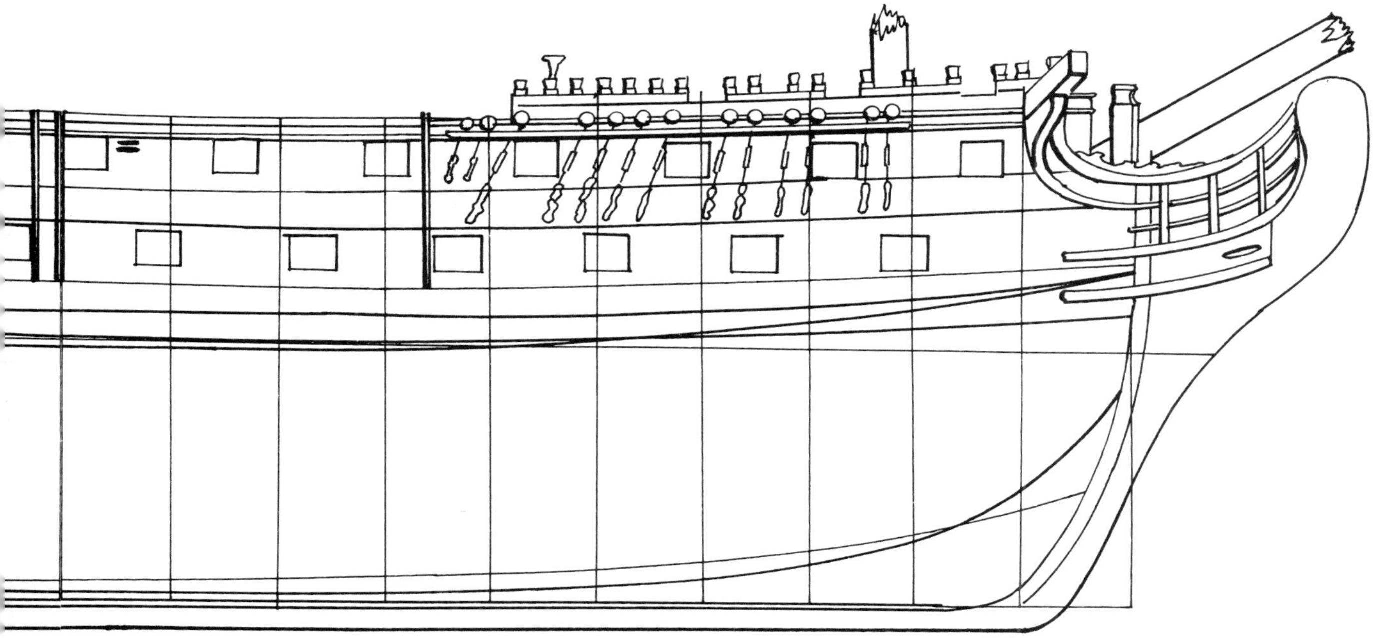

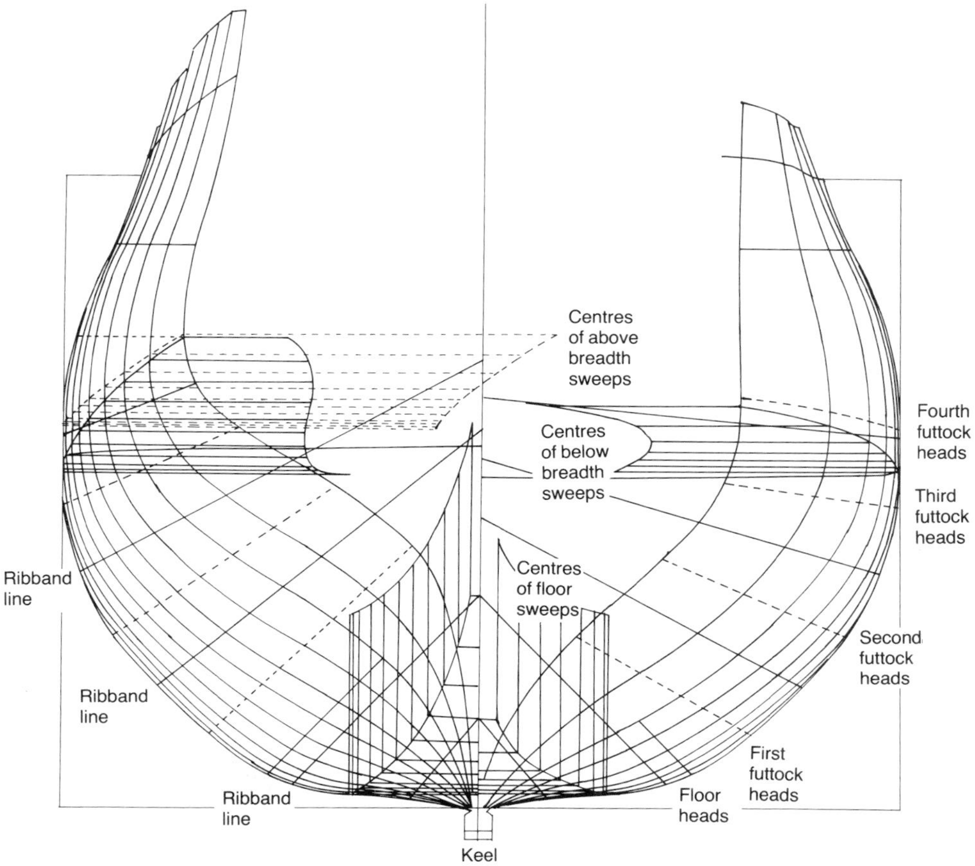

Centres of above breadth sweeps
Centres of below breadth sweeps
Fourth futtock heads
Third futtock heads
Ribband line
Centres of floor sweeps
Second futtock heads
Ribband line
First futtock heads
Ribband line
Floor heads
Keel

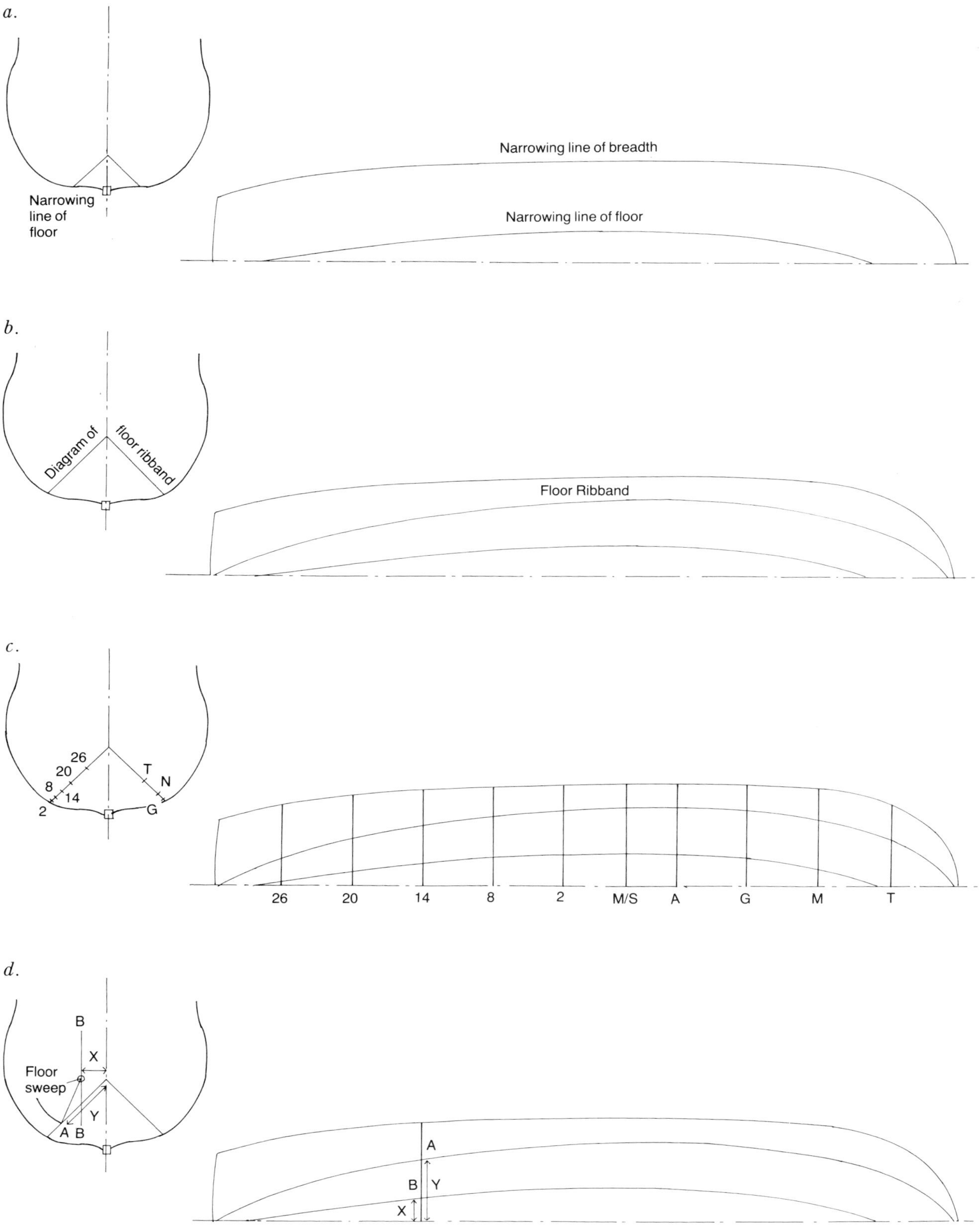
a.
Narrowing line of floor
Narrowing line of breadth
Narrowing line of floor
b.
Diagram of floor ribband
Floor Ribband
c.
26
20
8
14
2
T
N
G
26
20
14
8
2
M/S
A
G
M
T
d.
B
X
Floor sweep
Y
A B
A
B
Y
X

The use of rising lines
a. The old system, using narrowing lines of floor and breadth.

b. The diagonal of the floor ribband added. As drawn in the plan view, this is measured along the length of the diagonal in the body plan, whereas the narrowing line of floor is drawn showing its horizontal distance from the centreline.

c. Frame stations marked on the plan view, and transferred to the appropriate positions on the diagonals in the body plan.

d. Drawing the floor sweep on an individual frame. The point where it crosses the diagonal can be found from the body plan (y). The centre of the floor sweep is on the line B-B, as found from the narrowing line of floor, and the diameter is fixed. The centre can be found by drawing a sweep at centre A, and finding where that crosses B-B.

breadth under the planking in midships. Other key dimensions were the length of the keel, and the depth in hold – the measurement between the top of the keel and the underside of the lower gundeck. The number of guns was specified, including the calibres and numbers on each deck. For the rest, the ship designer relied on the accumulated experience of his profession, and a very limited scientific training.

The Sheer Draught

The designer began the sheer draught by drawing a straight line to represent the bottom of the keel of the ship. To this he added the stempost and sternpost. The sternpost was simply a straight piece of timber attached to the aftermost part of the keel. On British ships it tended to rise at an angle of about 10 degrees from the vertical, on French ships rather less; and, of course, the *Valiant* followed the latter practice. The stempost, which formed the basis of the bow of the ship, was very different. It formed a curve, often the arc of a circle, and this was usually tangential to the line of the keel.

The height of the lower deck above the keel was prescribed by the depth in hold, but it was also given a certain amount of sheer, curving upwards towards the bow and stern to allow water the drain to the centre of the ship. The other decks – upper gundeck, quarterdeck, forecastle and poop – had corresponding sheer and were positioned about 6ft apart to allow room for the movement of men. Above the sternpost, counters were drawn in the shape of segments of circles, and a straight line continued the shape of the stern upwards.

The size of the gunports had, by this time, been established by experience; they had to be large enough to allow the guns to be moved up and down and from side to side for aiming, but not so large that they would weaken the structure of the ship. Another factor was the distribution of the frames of the hull which, ideally, formed the sides of the gunports. The first gunport on each deck was placed close to the bow, though far enough aft to keep it away from the stern, leaving room for the quarter galleries which adorned that area. The gunports were spaced evenly along each deck. They were staggered with those on the deck above, to avoid creating areas of weakness in the hull.

Other parts of the ship indicated on the sheer plan included the wales. These were thicker pieces of planking running the length of the ship and with a greater sheer than the decks and gunports. The lower, or main wale, was drawn just under the lower deck gunports (though its greater curve might cause it to be cut by some of the ports near the bow and the stern); the channel wale was parallel to the main wale and was smaller and fitted between the ports of the lower and upper deck. As well as strengthening the hull, it provided a secure base for the chains which supported the standing rigging of the masts. Above the upper deck ports were several rails, also parallel to the lines of the wales, which were decorative as much as structural. The tops of the sides were formed by the gunwale.

The Midship Section

The draughtsman now turned to the lines of the ship which defined the complex underwater shape. He worked mainly with straight lines and circles, avoiding more subtle shapes such as ellipses and parabolae which the ill-educated shipwright of the time would not fully understand. The first task was to draw out the midship section of the ship, where it was widest and had its greatest area (though it was not always precisely in the middle of the length of the ship). He drew a

The mould loft at Chatham Dockyard.

rectangle bounded by the centreline of the ship, the greatest breadth within the planking, the line of the top of the keel, and the gunwale at the top of the side. Towards the lower part of the hull the shape would be relatively flat and this part was known as the floor. This was drawn as a straight line in the first instance, and continued out for about a third of the breadth. There it was joined by the first of the circles which formed the basic shape. This was the floor sweep, which was tangential to the straight line of the floor. The designer would also fix the height of the maximum breadth and drew this on the midship section. He decided the diameter of the breadth sweep, which would pass through this point, and drew it on the plan. Both the floor sweep and the breadth sweep were quite small in diameter, and were not intended to meet. To join them together, he used a larger curve known as the reconciling sweep, tangential to both the floor and the breadth sweep. This completed the shape of the hull below the height of the maximum breadth, though the floor itself was not always left flat being joined to the keel by means of a small concave curve.

The maximum breadth continued upwards for a short way, by means of a straight line known as the dead flat. After that the ship became narrower again, and this was known as tumble-home. It was believed that this helped to increase the stability of the ship as the weight of the guns was brought closer to the centreline (though this is doubtful according to modern theory), and that the narrower hull made the ship more difficult to board. Immediately above the maximum breadth, the hull was reduced by means of a convex curve, slightly smaller than the breadth sweep, and known as the above breadth sweep. This was joined by a large concave curve, the toptimber sweep, which carried the shape right up to the gunwale.

The Lines

The remaining sections of the hull were drawn out by means of a complex geometrical system, but all were based on the shape of the midship frame. In theory, each frame of the ship was drawn out using these sweeps, and some – the floor, reconciling and toptimber sweeps – retained the same diameter as used on the midship frame. The breadth sweep and the above breadth sweep, on the other hand, reduced in diameter towards the bow and stern, as this made it easier to produce a fair hull.

The positions of the breadth sweeps were determined by means of the lines of maximum breadth drawn on the sheer plan and the half breadth plan. These two lines, one in the vertical and one in the horizontal plane, determined the position of the maximum breadth for each individual frame. On the sheer draught, the line of maximum breadth was placed close to the lower wale, but was slightly less curved. In the horizontal plane, as shown on the half breadth plan, the line tapered gradually towards the bow and stern, from its maximum width at the midship frame. Close to the bow it formed an almost circular shape, and this was what gave the warship of the time its bluff bow. At the stern it formed a sharp angle with the transom so that the stern itself, at that level, was almost flat. Using the two lines of maximum breadth, the designer could determine the position of the breadth sweep for each individual frame, and draw it in as required.

Several methods were used to find the positions for the floor sweeps, but in the 1750s the most favoured was one known as the 'diagonal of floor ribband' system. A diagonal line was drawn in the body sections plan, starting at about the middle of the floor sweep of the midship frame, and angled upwards at about 45 degrees until it crossed the centreline of the ship. A corresponding line was drawn in the half breadth plan, having its widest point at the midships frame, and curving inwards towards the bow and stern. Another line was drawn inboard of this, and called the 'narrowing line of the floor'. This showed the positions of the centres of the floor sweeps in that plane. The draughtsman could find where the diagonal of the floor ribband crossed each individual frame from the half breadth plan, and transfer that distance to the body sections plan, giving a point through which the floor sweep must pass. He could find the distance of the centre of the floor sweep from the centreline by means of the narrowing line, and the diameter of the sweep was fixed, so he had enough data to draw it.

The shapes of the narrowing line of floor and the floor ribband lines were crucial to the underwater lines of the ship. There were no fixed rules for drawing them and most designers relied on their experience of past ships. If the lines were too 'full', the ship would be a poor sailer: if they were too sharp, the hull would not have sufficient displacement to support the guns. Deciding on the underwater shape was part of the long series of compromises which had to be

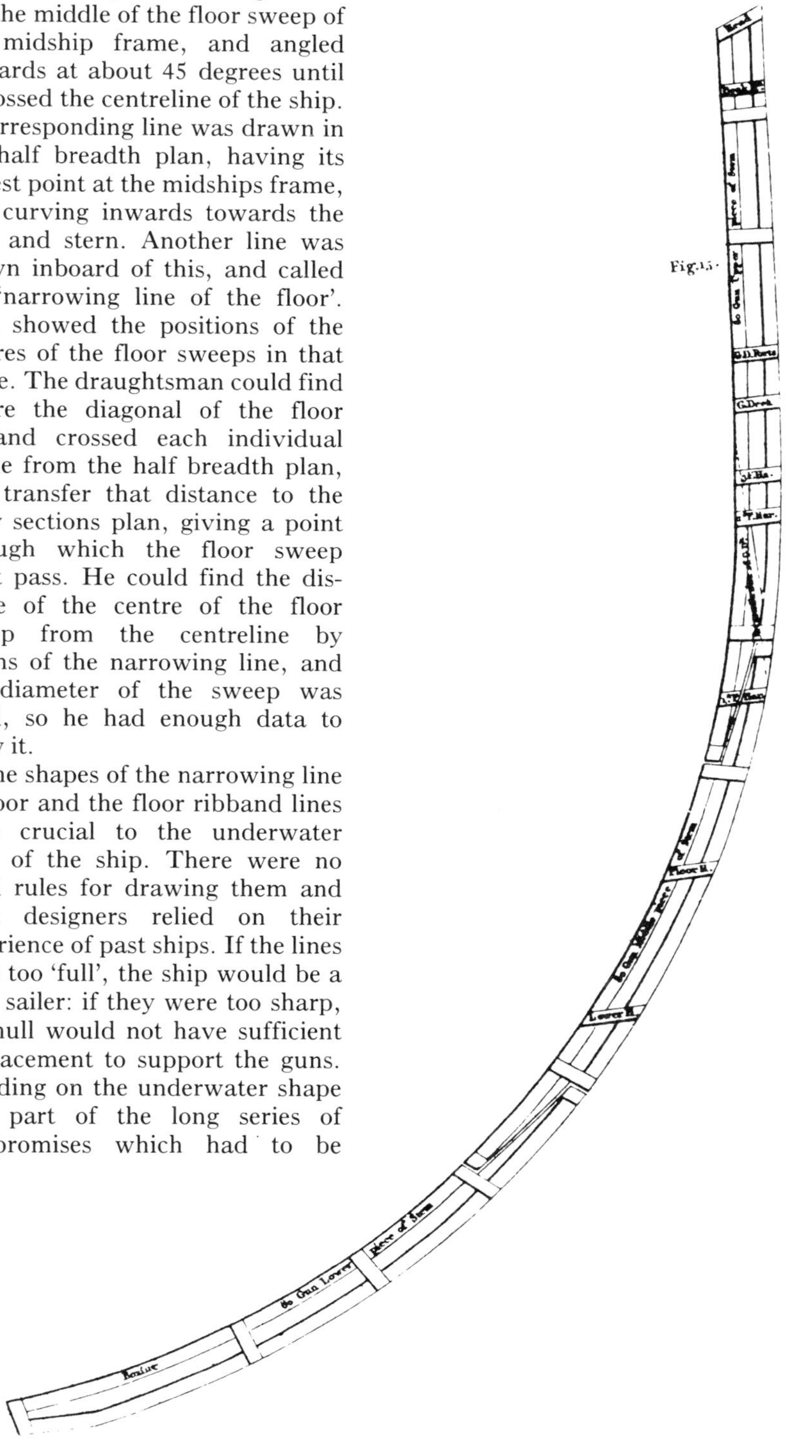

The mould for a stempost. (From Steel's Naval Architecture*)*

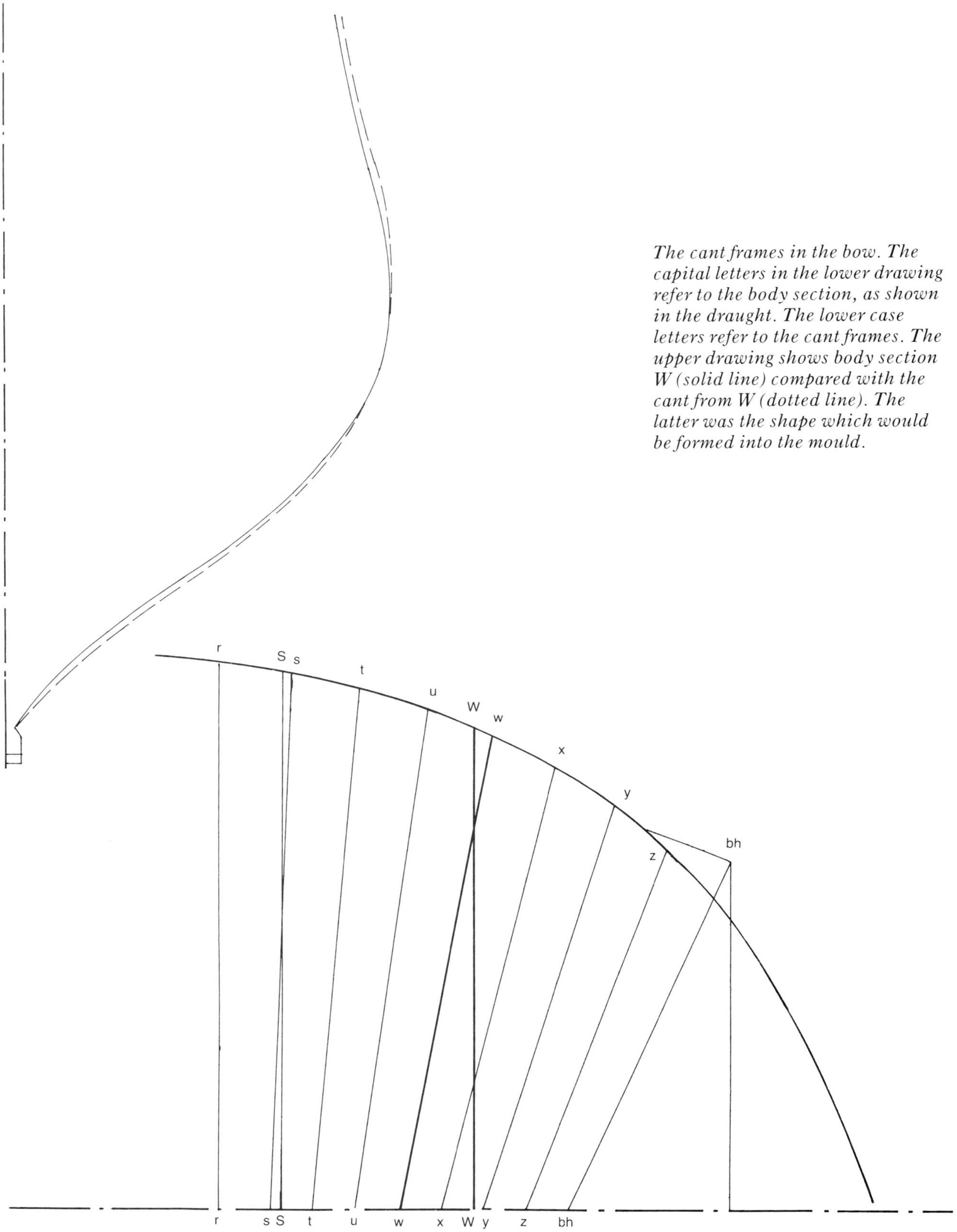

The cant frames in the bow. The capital letters in the lower drawing refer to the body section, as shown in the draught. The lower case letters refer to the cant frames. The upper drawing shows body section W (solid line) compared with the cant from W (dotted line). The latter was the shape which would be formed into the mould.

worked out in any ship design. The bow was kept bluff underwater and it tended to be sharper at the keel than at the waterline. The stern, below the level of the counter, was much sharper, partly to allow the water to reach the rudder and create a flow in which it could operate. After he had drawn in the individual frames, the draughtsman checked the fairness of his design by drawing in horizontal lines (known as waterlines) and diagonals on the sheer plan and body plans, and then translating them onto the half breadth plan, to make sure that the lines were smooth.

The shape above the maximum breadth was rather less crucial to a ship's qualities than the underwater section, and was somewhat simpler to draw. It was largely controlled by the position of the maximum breadth, which determined the positions of the above breadth sweeps. Another important factor was the toptimber line. This was drawn in the half breadth plan and the sheer plan, to show the shape of the tops of the sides. Given this information, it was simple to draw the toptimber sweeps for the individual frames, and the draught of the ship was complete.

Models

The Navy Board may have ordered a model to be made of the *Valiant* to go with the draught. This practice had been common a few years earlier, for an order of 1715 had decreed that a 'solid' or model should be made for each new or rebuilt ship. The actual purpose of such models, however, is obscure. They cannot have been much help in the design of the ships, since it often took as long to make a model as to build a ship; it is known that a model of the *Victory* of 1737 took more than four years to build.[1] Navy Board models, as they are known, were built in a peculiar style with the hull below the wales left unplanked and the framing thus exposed was constructed in a manner long obsolete for real ships, and which was very different from the actual framing used during the eighteenth century. No contemporary model of the *Valiant* has ever been found. Prints published in 1790 show the ship in Navy Board model form, but this might just have been a fancy of the draughtsman. The custom of making a model for every new ship had fallen into disuse under the pressure of war and it is not certain that one was ever made for the *Valiant*.

The Mould Loft

On its arrival at the building yard the paper plan, 1⁄48 of the size of the full-size ship, had to be converted into the three-dimensional man-of-war. The mould loft formed the link between the stages of drawing and building; full size patterns, or 'moulds', were made for the individual frames and for the separate pieces of timber which formed them. In 1757 the mould loft at Chatham was situated above the new mast house, begun in 1753, and it occupied a large open space, 155ft by 70ft which was largely uninterrupted by pillars and other supports and was well lit from windows in the roof.

Moulds were made from fir about 3⁄4in thick; the drawing was done by shipwrights while joiners actually cut out the moulds. Each was made of a light framework so that curved pieces of timber, sawn to shape, represented the inner and outer edges of the frames, and they were joined by battens nailed between them. The draughtsmanship was quite simple in the case of the 'square' frames which would occupy the central part of the ship, for these could simply be scaled up from the draught. The timbers near the bow and stern were set at an angle; their actual shapes were not shown in the draught, so the shipwrights had to calculate their precise shapes. When they were finished, the moulds were taken down to the timber piles and building slips of the dockyard, and used to select and cut the pieces of timber which would be used for the frames. The building was ready to begin.

4

Chatham Dockyard and its Men

Chatham Dockyard in 1752. (National Maritime Museum)

Chatham Dockyard was a large industrial complex in the eighteenth century, even though it was no longer the largest dockyard in England, as it had been a century earlier. Britain's main enemies were now the French and Spanish, rather than the Dutch, so the south coast yards at Portsmouth and Plymouth had become pre-eminent. Nevertheless, Chatham was an enormous enterprise by the standards of the time – far bigger than any civil factory in the age before the industrial revolution. It employed more than 1,700 men at a time when industry was largely rural and small scale.

The Dockyard Facilities

Chatham Dockyard was, and is, sited on the east bank of the River Medway, about ten miles from the mouth of the river. Its site was not ideal, for the river was narrow and winding, and it often took weeks to get ships up and down the river when the winds were unfavourable. The yard did not expand greatly in the eighteenth century but many of its buildings were quite new in 1757. At the centre of the yard, and giving it its main *raison d'etre*, were four dry docks which were intended for the repair and cleaning of ships. In the mid-eighteenth century, however, the most urgent repairs were done in the south coast yards where the ships could be kept closer to the scene of the battle. As a result there was some spare dry-dock capacity at Chatham and this was sometimes used for shipbuilding as well as repair even though it would result in a dock being occupied for two years or more.

More conventionally ships were built on building slips, of which Chatham had three. Besides the dry docks and slips there were about forty saw pits, where logs could be cut to shape. There were also large open spaces in the yard, where wood could be stored in the open in its cut or uncut state, and allowed to season. There were mast houses and boat houses, and a great forge where anchors could be made. Rope was made in the ropery at the southern end of the yard (on almost the same site as the ropery in use today), and sails were produced in the sail loft. Supplies were largely brought into the yard by ship; and cranes, operated by treadmill or capstan, were placed along the river bank for unloading. There were great storehouses by the side of the river, where hundreds of items needed to keep a man-of-war afloat and operational – nails, lanterns, tar, hemp, compasses, hammocks, sails and much else – were kept in relative security. There was a small office block from which the works of the yard were directed.

The River Medway

As well as its facilities on dry land, the dockyard had responsibility for the ships and hulks moored in the River Medway, over a stretch of about five miles from Rochester Bridge to Gillingham Reach. In peacetime, the river was filled with ships laid up 'in ordinary'. These

A ship in dry dock at Chatham, c.1790, painted by John Williams who was Clerk of the Cheque at the time.

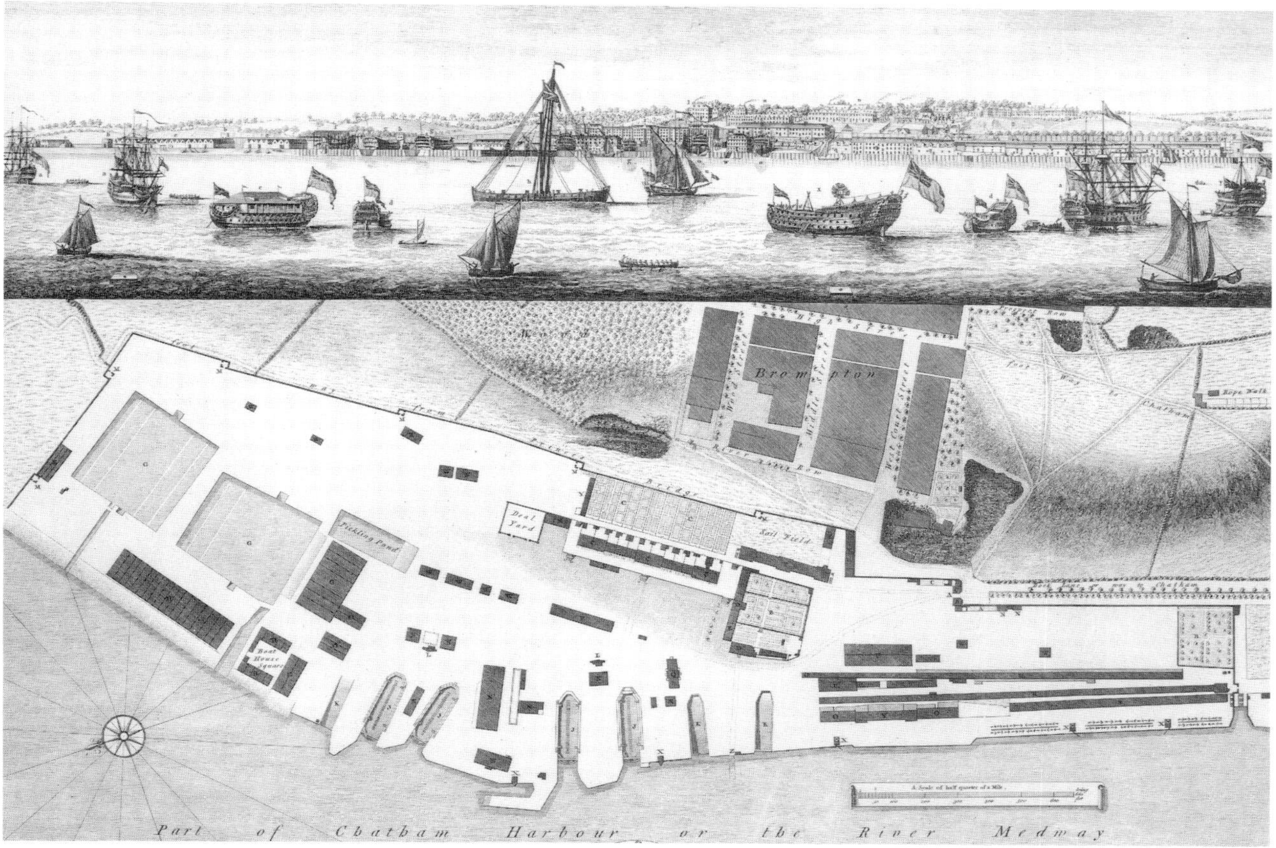
Brompton
Middle Street
Rope Walk
Deal Yard
Sail Field
Pickling Pond
Boat House Square
Part of Chatham Harbour or the River Medway

Bridge Reach.
Limehouse Reach.
Chatham Reach.
Rochester
and
Chatham.

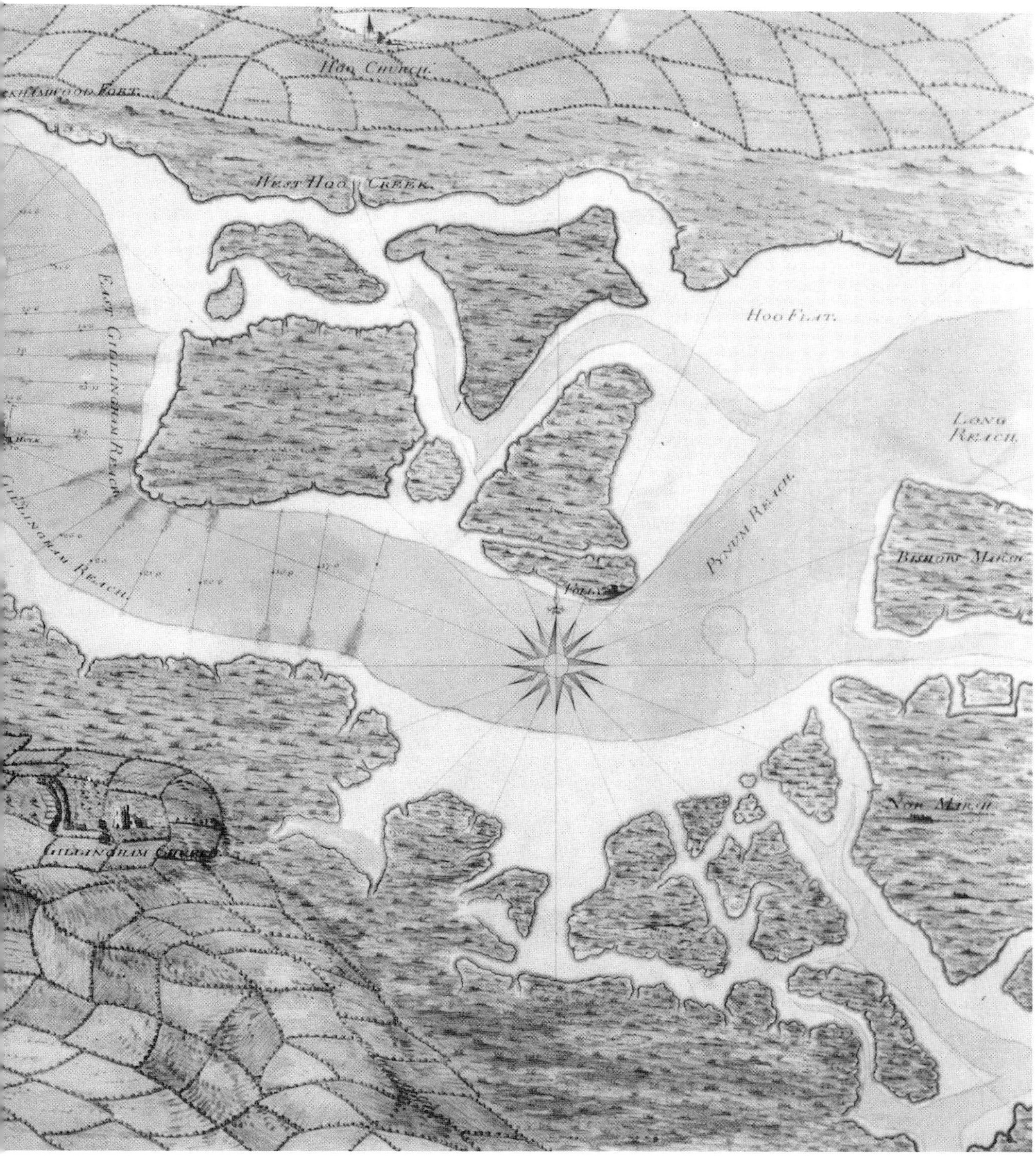

The River Medway in 1771, showing the dockyard, victualling yard and ordnance yard, with the fortifications and the moorings in the river. (British Library)

had their guns, masts and stores taken out, and were looked after by the gunner, boatswain and carpenter of each, assisted by servants and labourers. 'Guardships' were in a greater state of readiness, with almost a full complement of officers, most of the guns and rigging on board, and a small crew which could be rapidly expanded if the need arose. In wartime the ships in ordinary were mostly put into commission, but their place at the moorings was often taken by others – ships awaiting repair, or hulks which no longer put to sea but served various purposes. 'Sheer hulks' were specialised vessels fitted with legs or 'sheers' to lift lower masts into ships. 'Receiving ships' held men, often taken by the press gang, who were awaiting allocation to a ship in commission. Hospital ships were moored in the river, along with 'lazarettos', or quarantine ships, used to accommodate men suspected of having infectious diseases. All these vessels lay in the centre of the river, fixed bow and stern to great mooring buoys.

In addition to these moored craft, there were many small vessels which plied about the river. Rowing and sailing boats took men out to the hulks, while barges constantly brought in supplies of timber, tar and many other items. The yard officers had their own yacht to take them from place to place, and each had his wherry in which he was rowed about by a crew of watermen.

All the officers of the yard, from the Commissioner down to the Assistant Master Shipwrights, were allowed houses within the site. The Commissioner's House was a notable building, erected in 1703 and decorated with a ceiling painting taken from an old 100-gun ship, the *Royal Sovereign*. The other

The Commissioner's House, built in 1703, and now the oldest intact naval building in the country.

The main gatehouse at Chatham, built in 1720.

houses were arranged in a terrace overlooking the yard. For security against theft or sabotage a high wall surrounded the yard, and guard towers were placed along its length. An imposing gatehouse, built in 1720, controlled the main entrance. For defence against more substantial forces, such as an invading army, the 'lines' were being built around the yard during the 1750s. These consisted of entrenchments, using the natural defences offered by the hills at Brompton, to the east of the dockyard. For more long-range defence, the Medway was protected by forts at Sheerness at the entrance to the river, and by twin forts on each side of the river at Gillingham and Cockham Wood.

The Ordnance and Victualling Boards

The dockyard itself was only the central part of a great naval base. The armaments for ships were supplied by the Board of Ordnance, which was a separate government department, independent of both the Admiralty and the Navy Board. The main armament depot, situated along the river front to the south of the dockyard, supplied and stored the guns for the fleet. Gunpowder was stored on the other side of the river to the north of the dockyard, in Upnor Castle.

The Victualling Board supplied the ships with food and certain other stores, under the direction of the Board of Admiralty. Its premises were situated in the town of Chatham, about a mile away from the dockyard. They were quite small, and supplied only the ships in the River Medway itself; ships preparing for longer voyages usually anchored at the Nore, in the Thames estuary, and took on supplies sent down from Deptford. The Chatham victualling yard had

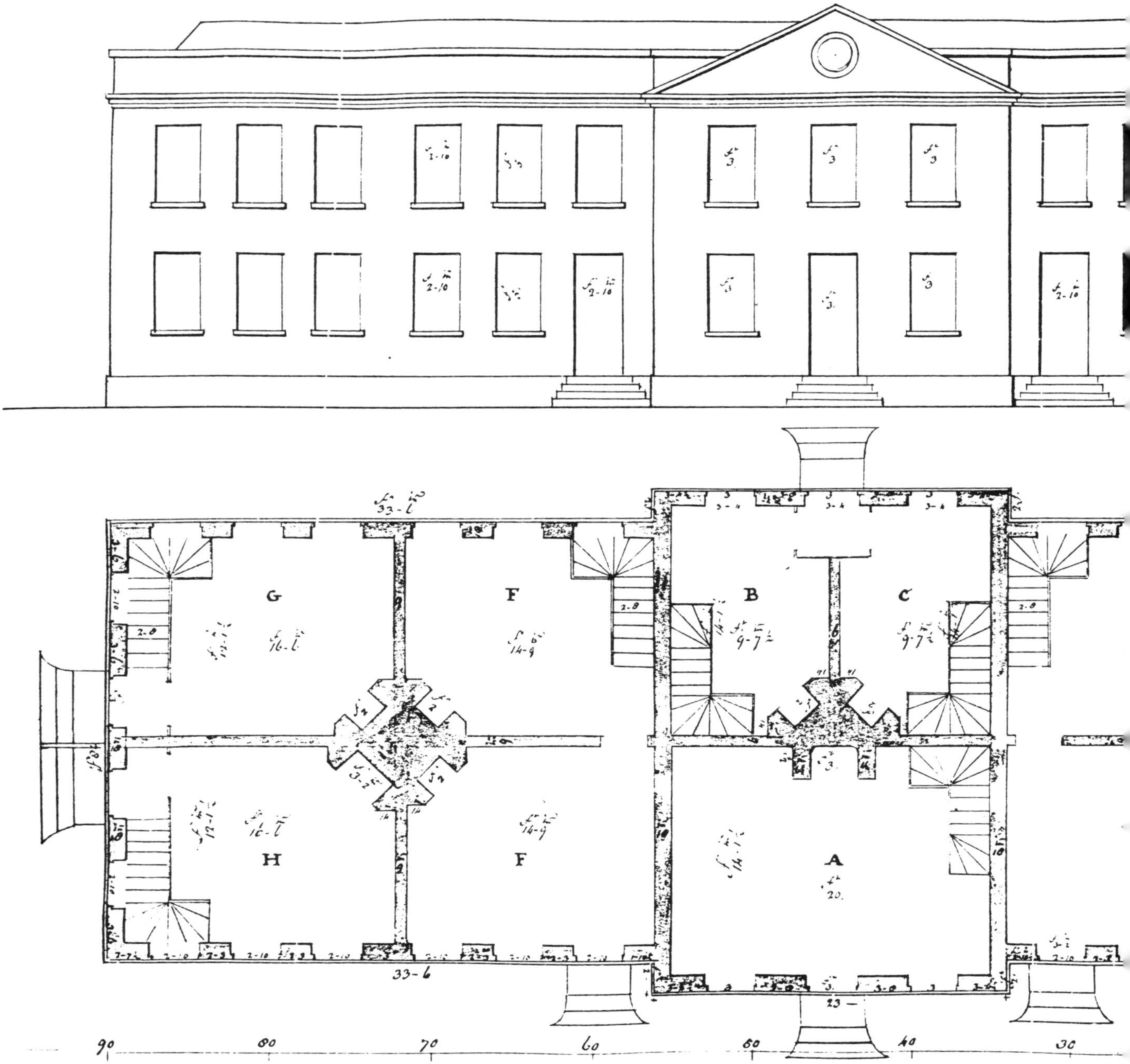
G
F
B
C
H
F
A
33-6

The office building at Chatham (since demolished), showing how each officer had his own entrance, with no internal communication. (National Maritime Museum)

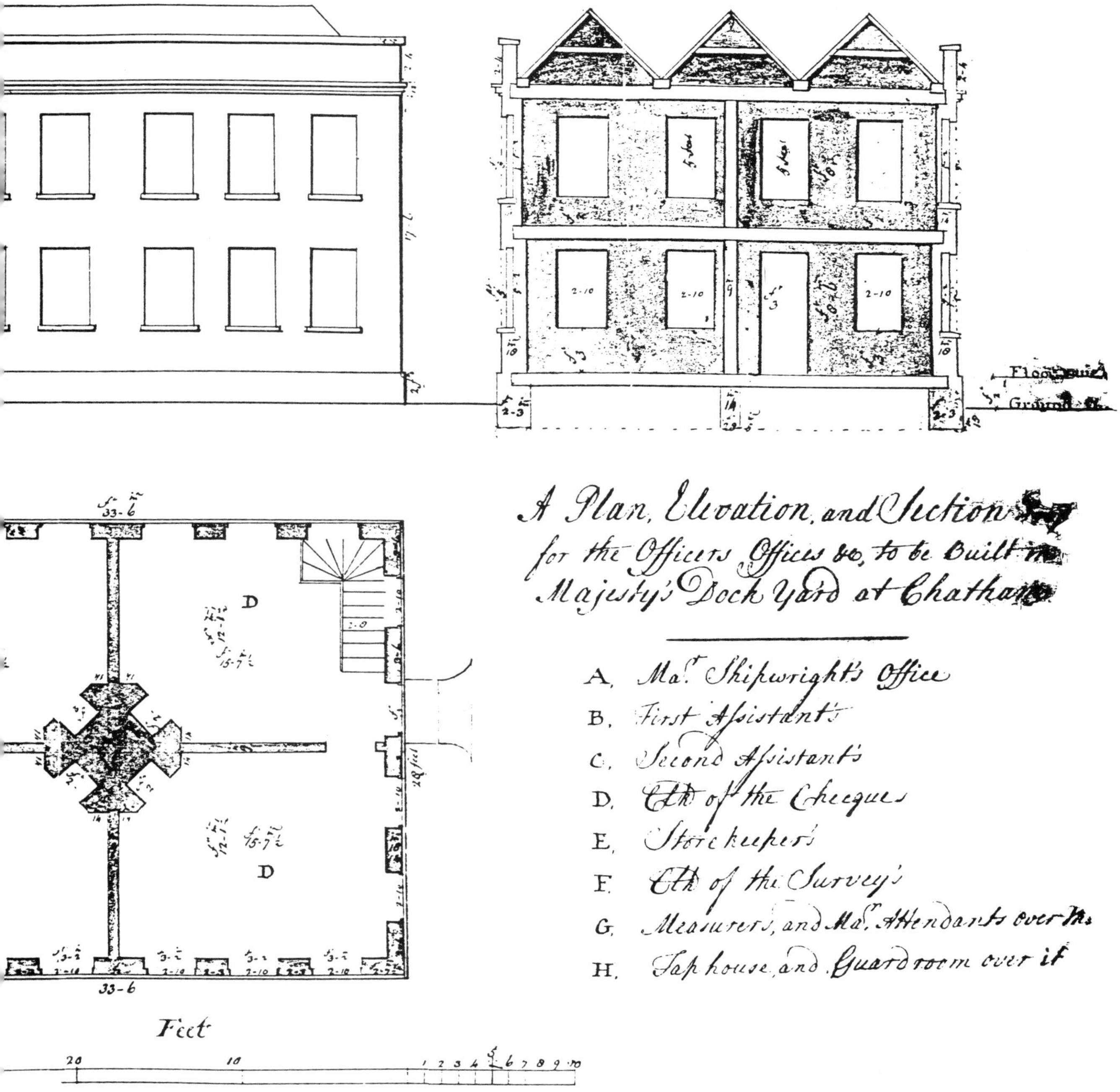

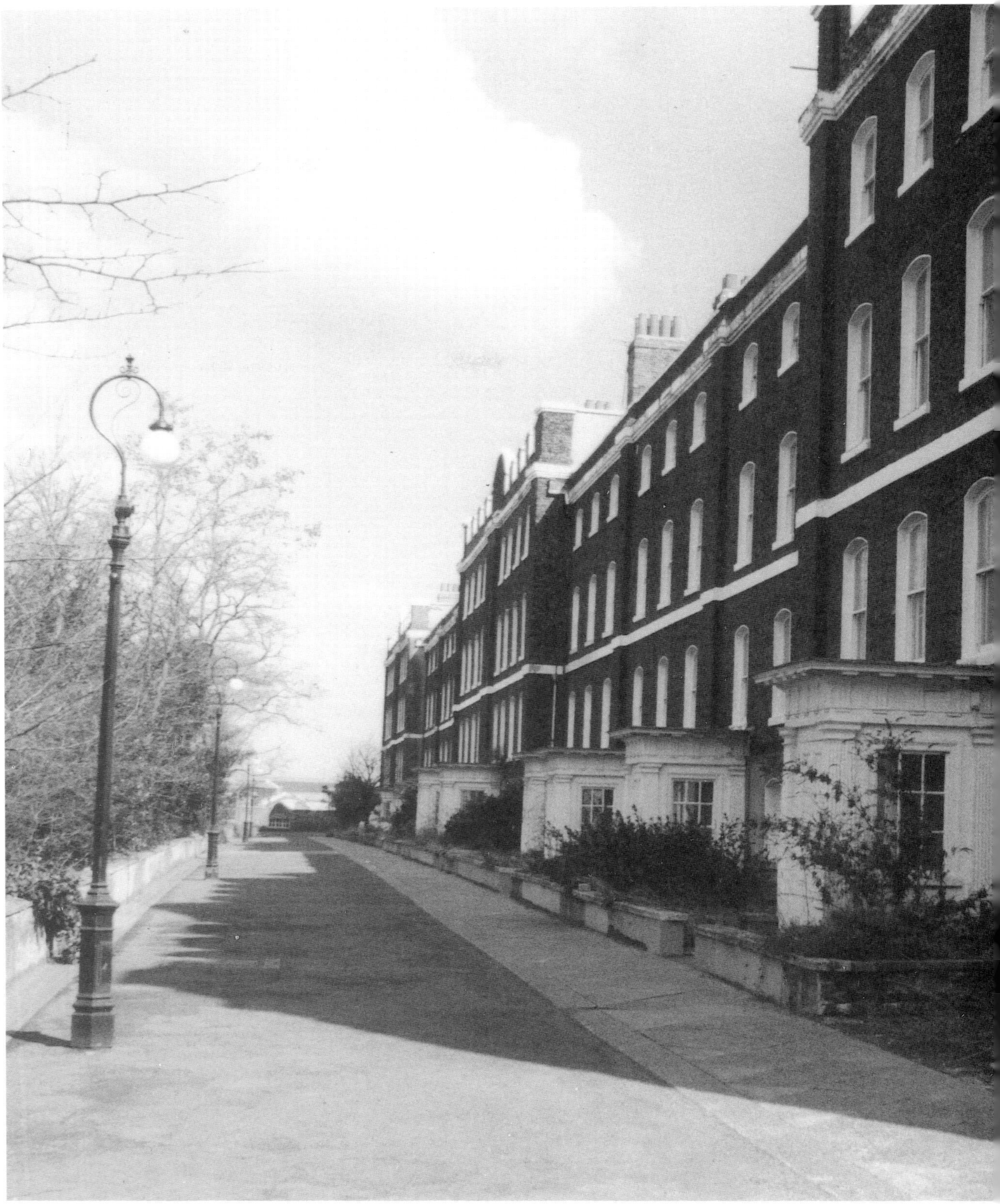

The officers' terrace in Chatham Dockyard where the senior officials of the yard lived.

storehouses and offices, and a slaughterhouse for cattle, but no facilities for making biscuits or beer, as were provided at the larger victualling bases.

The Town of Chatham

The town of Chatham provided the accommodation for most of the workers in the dockyard and the other naval and military installations. It had expanded from a small fishing village during the previous two centuries but it had not grown into an attractive town. Visitors invariably noticed the ancient city of Rochester with its medieval castle, cathedral and bridge, and commented on the great dockyard at Chatham; but they paid little attention to the town of Chatham itself, situated between the two. One exception was Edward Hasted, who described it as 'a long, narrow, disagreeable, ill-built town'.[1] The suburb of Brompton, in the hills above the dockyard, offered slightly better accommodation, and provided homes for some of the middle ranking officials of the navy.

Dockyard Administration

Nominally the head of the dockyard was the Commissioner. He was technically a member of the Navy Board in London though he rarely had an opportunity to attend its meetings. In practice his power was rather limited as the workforce of the yard was more directly under the control of the other officers, while ships in commission were controlled by the port admiral. A dockyard commissioner was usually an experienced naval captain, and Thomas Cooper, the Commissioner at Chatham at the time, conformed to this pattern. He had commanded a ship of the line at the

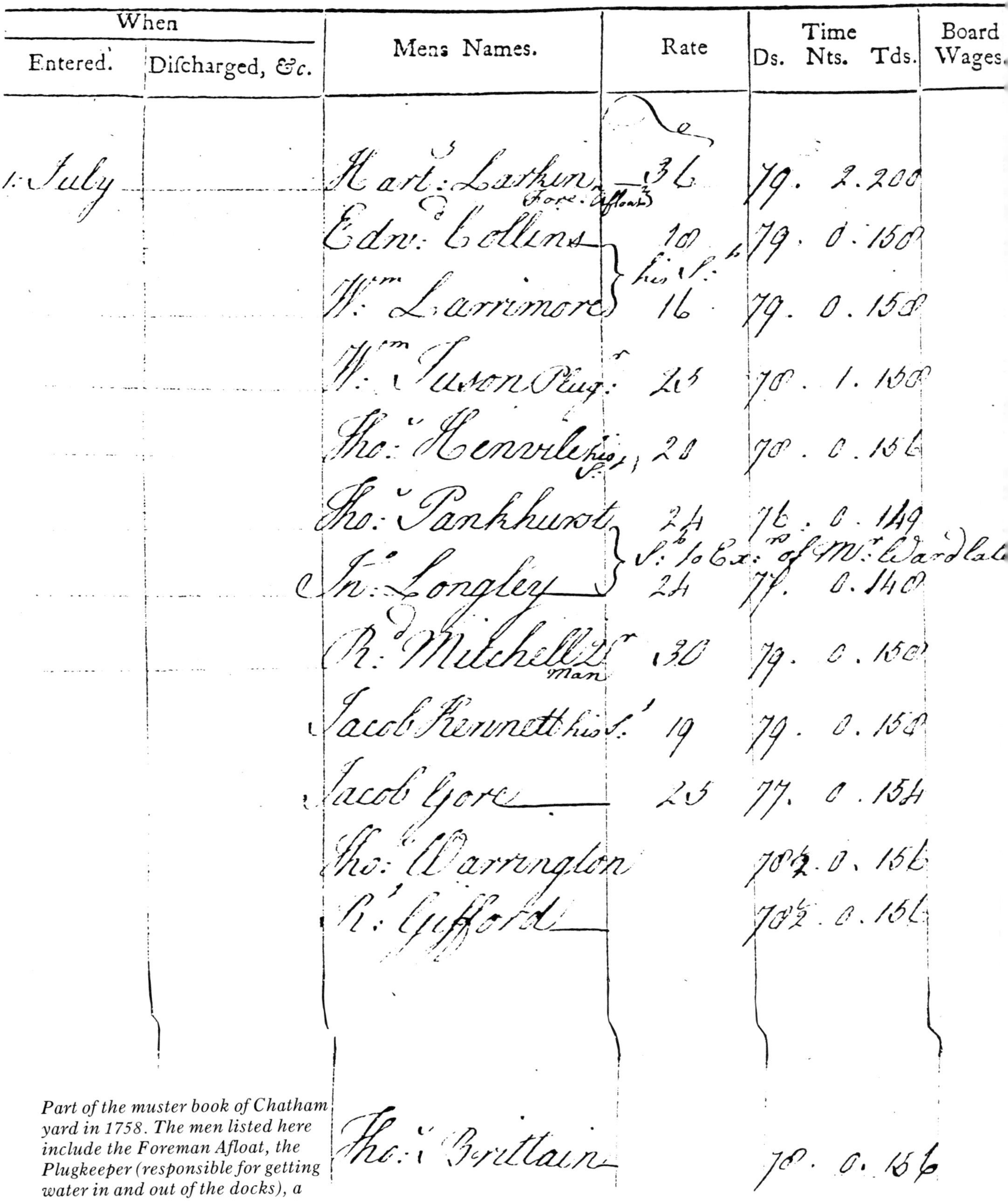

When		Mens Names.	Rate	Time			Board Wages.
Entered.	Discharged, &c.			Ds.	Nts.	Tds.	
1 July		Hart: Larkin Fore: Afloat	36	79.	2.	200	
		Edw: Collins his S.	10	79.	0.	150	
		Wm Larrimore	16	79.	0.	150	
		Wm Juson Plugr	25	70.	1.	150	
		Tho: Henville his S.	20	70.	0.	156	
		Tho: Pankhurst	24	76.	0.	149	
		S. to 6 Ex: of Mr Wardla...					
		Jn: Longley	24	77.	0.	140	
		Rd Mitchell Qr Man	30	79.	0.	150	
		Jacob Kennett his S.	19	79.	0.	150	
		Jacob Gore	25	77.	0.	154	
		Tho: Warrington		78½.	0.	156	
		R: Gifford		78½.	0.	156	
		Tho: Brittain		78.	0.	156	

Part of the muster book of Chatham yard in 1758. The men listed here include the Foreman Afloat, the Plugkeeper (responsible for getting water in and out of the docks), a quarterman and several servants. (Public Record Office)

Neglect, &c.	Full Wages	Lodging.	Solv. Colu.	Neat Wages.		To whom Paid, &c.
	18 . 0 . 0	2 . 0	/			
	0 . 11 . 2	2 . 0	/	34 . 10 . 0		
	7 . 4 . 10	2 . 0	/			
	13 . 3 . 4	2 . 0	/			
	9 . 15 . 0	2 . 0	/	23 . 2 . 6		
	11 . 18 . 11	2 . 7	/			
mast: Shipwright	12 . 0 . 4	2 . 0	/	24 . 3 . 6		
	14 . 16 . 10	2 . 0	/			
	9 . 4 . 4	2 . 0	/	24 . 4 . 6		
	12 . 16 . 0	2 . 0	/	12 . 18 . 6		
	13 . 1 . 0	2 . 0	/	13 . 3 . 0		
	13 . 1 . 0	2 . 0	/	13 . 3 . 0		
	13 . 0 . 0	2 . 0	/	13 . 2 . 0		

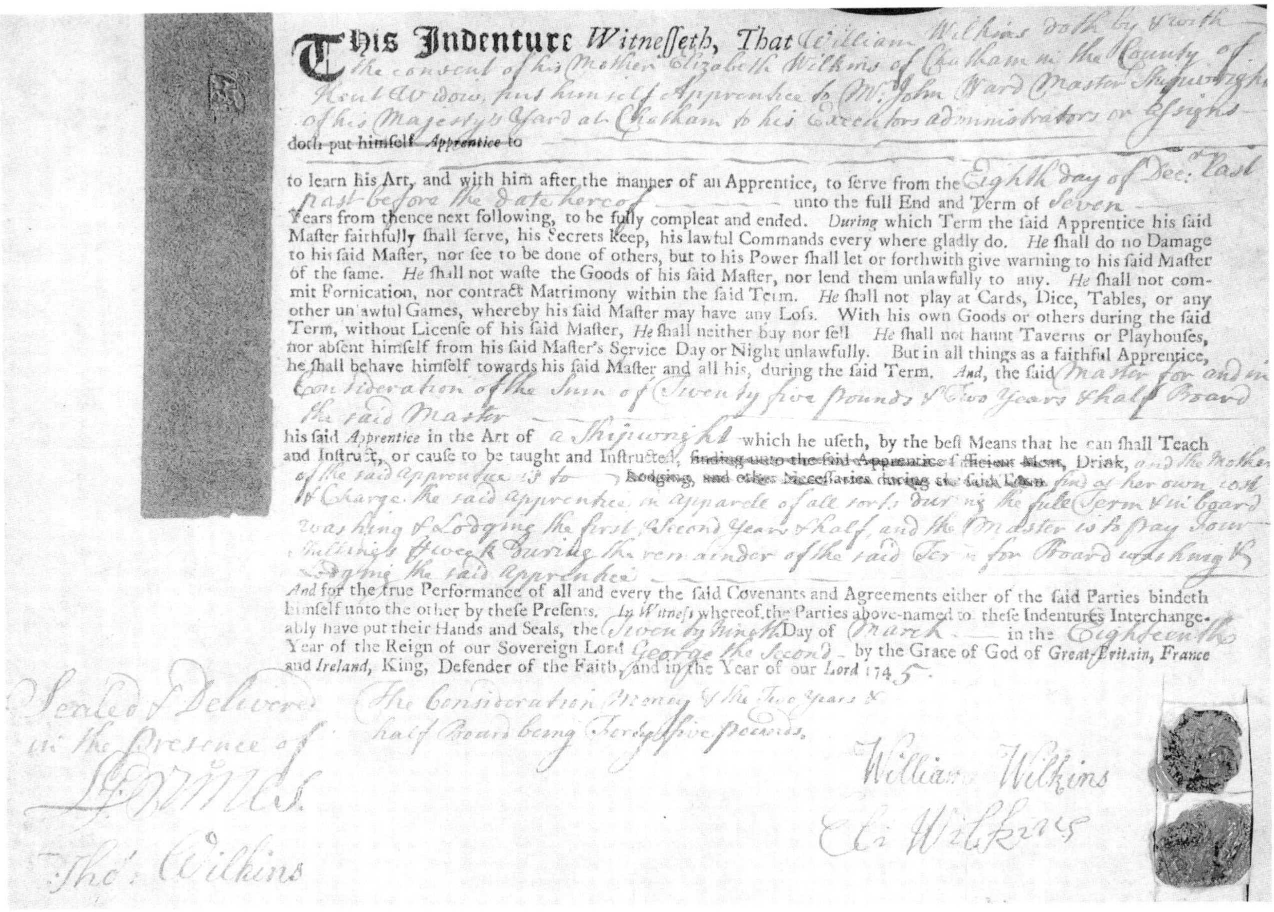

This Indenture Witnesseth, That William Wilkins doth by & with the consent of his Mother Elizabeth Wilkins of Chatham in the County of Kent Widow, put himself Apprentice to Mr. John Ward Master Shipwright of his Majesty's Yard at Chatham to his Executors administrators or Assigns ~~doth put himself Apprentice to~~

to learn his Art, and with him after the manner of an Apprentice, to serve from the Eighth day of Dec: last past before the date hereof unto the full End and Term of Seven Years from thence next following, to be fully compleat and ended. During which Term the said Apprentice his said Master faithfully shall serve, his Secrets keep, his lawful Commands every where gladly do. He shall do no Damage to his said Master, nor see to be done of others, but to his Power shall let or forthwith give warning to his said Master of the same. He shall not waste the Goods of his said Master, nor lend them unlawfully to any. He shall not commit Fornication, nor contract Matrimony within the said Term. He shall not play at Cards, Dice, Tables, or any other unlawful Games, whereby his said Master may have any Loss. With his own Goods or others during the said Term, without License of his said Master, He shall neither buy nor sell. He shall not haunt Taverns or Playhouses, nor absent himself from his said Master's Service Day or Night unlawfully. But in all things as a faithful Apprentice, he shall behave himself towards his said Master and all his, during the said Term. And, the said Master for and in consideration of the Sum of Twenty five pounds & Two Years & half Board the said Master his said Apprentice in the Art of a Shipwright which he useth, by the best Means that he can shall Teach and Instruct, or cause to be taught and Instructed, ~~finding unto the said Apprentice sufficient Meat~~, Drink, and the Mother of the said apprentice is to ~~Lodging, and other Necessaries during the said Term~~ find at her own cost & Charge the said apprentice in apparel of all sorts during the full Term & in board washing & Lodging the first, Second Years & half, and the Master is to pay Four Shillings a Week during the remainder of the said Term for Board washing & Lodging the said apprentice

And for the true Performance of all and every the said Covenants and Agreements either of the said Parties bindeth himself unto the other by these Presents. In Witness whereof the Parties above-named to these Indentures Interchangeably have put their Hands and Seals, the Seven & Twentieth Day of March in the Eighteenth Year of the Reign of our Sovereign Lord George the Second by the Grace of God of Great-Britain, France and Ireland, King, Defender of the Faith, and in the Year of our Lord 1745.

Sealed & Delivered in the Presence of [illegible] Thos. Wilkins

The Consideration Money & the Two Years & half Board being Forty five pounds.

William Wilkins
Eli Wilkins

The indenture of William Wilkins who began his apprenticeship at Chatham Dockyard in 1745. (National Maritime Museum)

unsuccessful Battle of Toulon in 1744, and had been court-martialled as a result. He was dismissed from the service on a technicality but soon reinstated and he then served as a member of the Victualling Board and the Navy Board, before being appointed to Chatham in 1755.

Since shipbuilding and repair were the main functions of the dockyard, the Master Shipwright was a key figure. The post was held by John Lock, who had begun his career at Plymouth where his father, Peirson Lock, was Master Shipwright.[2] He had come to Chatham in 1755, after experience as Assistant Master Shipwright at Portsmouth and Plymouth, and Master Shipwright at the latter yard. He was helped by several assistants, generally younger men who hoped to become Master Shipwrights themselves one day, or even rise to become Surveyor of the Navy. A Master Shipwright normally spent his whole career in the royal dockyards, beginning as an apprentice to a senior figure in the yard. Unlike the common workmen who simply cut and assembled timber, he would learn the secrets of ship design in the mould loft and drawing office; in effect, he was a naval architect though the term was not yet in common use, and officially he belonged to the same profession as the humbler shipwrights under his command.

Besides the Commissioner and the Master Shipwrights, there were four other principal officers of the yard, in addition to the Clerk of the Ropeyard who was almost independent of the others. The Master Attendant was an experienced seaman and was responsible for the ships laid up in ordinary at the yard, for their moorings, and for navigation, pilotage and buoyage in the area. The other senior officers, the Storekeeper, Clerk of the Survey and Clerk of the Cheque, had overlapping responsibilities, mainly to do with accounting, storekeeping and administration. When these posts had been created many years earlier, it had been hoped that each would keep a check on the other. The officers had their working accommodation in a block, long since demolished, in the centre of the yard. It is notable that each officer had his own entrance to the building, and there was no internal communication between the separate departments.

The Shipwrights

Chatham dockyards employed 1,751 men (excluding officers and clerks) on 6 October 1758. Of these, the shipwrights were easily the largest group, numbering 725. A shipwright was equally at home in the building or repair of ships and, indeed, many left the dockyard to become ship's carpenters responsible for the maintenance of ships afloat. Every shipwright had to enter the trade by means of apprenticeship, starting at the age of about 15 and being bound to his master for seven years. During that time he was forbidden to 'commit fornication, nor contract matrimony . . . play at cards, dice or any other unlawful games . . . haunt taverns or playhouses'.[3] His master took his wages and undertook to train him and provide him with food and clothing. The apprentice was indentured to an individual workman, not to the yard or the navy as a whole. Well connected young men might be apprenticed to a Master Shipwright or his Assistant, but for those of humbler background there were only two alternatives within the dockyard – to serve under a ship's carpenter or an ordinary shipwright. Every carpenter was allowed a boy or two according to the size of his ship, but for ordinary workmen in the yard it was a special privilege, awarded for good work and faithful service, or perhaps as a result of bribery in some cases.

A man who attempted to enter the trade by irregular means might find himself in severe danger from his workmates. At Portsmouth in 1741 a man had been detected with false indentures, and 'the shipwrights surrounded him, put a piece of quarter between the legs, took him upon their shoulders, carried him just without the gates, then set him down, gave three shouts . . . and returned to their duty'.[4] The shipwright had great pride in his trade. As well as the skills of building the hull of a ship, he could usually turn his hand to mastmaking and boatbuilding, so there was no separate establishment of such workers within the yard.

Under the Master Shipwright were

The rural shipbuilding site at Bucklers Hard, as it is today.

Blackwall yard in the 1780s with several 74-gun ships under construction. (National Maritime Museum)

several foremen, some responsible for new work, others for repairs. The Master Mastmaker and Master Boatbuilder were on a par with the foremen. Under each foreman were several quartermen and each of these headed a gang of about ten men.

Other Trades

Of the other tradesmen within the yard, some had no direct connection with shipbuilding. There were eighty house carpenters, twenty bricklayers, and two plumbers working on the maintenance of the buildings. Most of the other skilled workers made some contribution to the building of the ships; for instance, there were forty-three pairs of sawyers, preparing timber for the shipwrights. The caulkers who worked on the finished ship, making it watertight, numbered seventy-six and they were assisted by thirty oakum boys. There were forty-four sailmakers, seventy-three riggers, seventy-four smiths, two blockmakers and two trenail mooters. The ropeworks employed a further 147 tradesmen. All these men had served apprenticeships in roughly the same way as the shipwrights.

There was also a large number of unskilled labourers. Some were employed solely on specific tasks such as the sixteen bricklayer's labourers and thirty-two rigger's labourers, who brought the materials to their masters. There were 191 yard labourers on more general duties, and seventy-one 'scavelmen', who cleaned out the mud from docks, and did other dredging work.[5]

Transport within the yard was carried out by nine teams of horses.

Working Conditions

The normal working day was from 6am to 6pm, with long breaks for meals. The dockyard bell rang to signify the beginning and end of work, and the men were mustered several times during the day by sending officials round to check that they were actually on duty. Sawyers were paid on piece work, according to the amount of timber cut; other men were paid 'by the day'. Overtime was often worked, especially in wartime; a 'tide' lasted 1½ hours, while a 'night', of five extra hours, earned a full day's pay. The men were paid quarterly, in arrears, so it was possible to work up to six months before getting any wages and so, inevitably, the workmen established systems of credit with local landlords and shopkeepers to survive during this period.

As in many large industrial organisations, the workmen were often accused of laziness, corruption and truculence. One constant bone of

contention was the system of 'chips'. Each shipwright was entitled to take home pieces of timber which were too small for shipbuilding. It was said that men either destroyed valuable large pieces to make them into chips, or took out large pieces with the connivance of the warders on the gate. Industrial relations were often bad, and there were several strikes in the course of the eighteenth century – most recently in 1756, when the men refused to be searched for oversize chips, and stormed the gates.[6] The work was labour intensive and often boring, and sometimes there were severe shortages of labour or materials, and work was held up. Nevertheless, Chatham Dockyard carried out hundreds of repairs of ships, and during the Seven Years War (1757–63) the yard produced sixteen warships, including four of 90 guns, four of 74, and three of 64 or 70 guns. The yard's most famous product in these years was the *Victory*, begun in 1759 but not launched until after the war. Much of the yard of 1757–59 survives today, and can be visited.

Other Yards

Chatham Dockyard was only one of many places where warships were built in the eighteenth century. There were five other dockyards, situated on the River Thames or the south coast. Working conditions were similar in all the naval dockyards, though the emphasis varied from place to place. Deptford and Woolwich were both far up the Thames, and far from the scene of action. Both concentrated on shipbuilding and major repairs, rather than on urgent services to the fleet. Ships built there were not fitted with the guns and stores until they had sailed some way down the river because the water was so shallow. Sheerness, situated at the mouth of the Medway, where it joins the Thames Estuary, specialised in small ships such as frigates, rather than ships of the line. On the south coast, the main bases were at Portsmouth and Plymouth and despite their commitments to fitting out and servicing the active fleet, both yards built many new ships during the eighteenth century.

In wartime, most of the navy's ships were built in private shipyards so that the facilities of the Royal Dockyards would not be over-strained. The Admiralty liked to keep the building under its scrutiny, so contracts were only awarded to yards which were close to the government yards – on the Thames, Medway, in the Solent and in Plymouth Sound. Builders were invited to submit tenders, and the lowest one was accepted. A typical contract, for a 74 of the *Dublin* class, was signed with Thomas West of Deptford in 1755. He was to be paid £17.3.6 per ton, so that the whole ship would earn him about £26,000. Payment came in eight parts, beginning with the signing of the contract, and then as various stages in construction were reached.[7] A naval overseer, usually a dockyard foreman, was appointed to see that no shoddy workmanship or defective materials went into the ship.

Workers in the private yards tended to be better paid than on the Royal Dockyards, especially when a war created a shortage of labour; but they were taken on for a particular contract only and had less security of employment. Some of the largest yards were in the Thames, quite close to London, and the Blackwall yard was the largest in the country. Other shipyards were established on more rural sites – one, which still survives, was situated at Bucklers Hard in Hampshire.

5

Finding the Timber

Timber was the main material for shipbuilding and the most important ingredient of naval power (except perhaps for skilled seamen). The navy expanded greatly during the Seven Years War, and afterwards in 1768 the Navy Board complained that 'The counties which used to afford the largest supplies [of oak timber] are so far exhausted, that it is with difficulty they can procure one thousand loads for naval service in a circuit of many miles'.[1] (A load was 50cu ft of timber, equivalent to a substantial tree.) In 1771 a House of Commons committee was told by a timber merchant that suitable timber in the main oak producing counties – Sussex, Kent, Berkshire, part of Hampshire, Essex, Suffolk and part of Norfolk – had almost all been cut down in the last 30 years.[2]

Shortage of timber could limit the overall size of the navy, while the properties of any timber itself could limit the size of the individual ships. Because timber could only be grown to a certain size, ships could not be made much bigger using the construction methods in use in the eighteenth century.

The *Valiant* was built before the timber problem reached a state of crisis. The English forests had been depleted a century before, by Oliver Cromwell, when the estates of Royalist landowners had been pillaged to build a great navy. As a result, John Evelyn had written a book entitled *Sylva*, in which he had urged landowners to plant trees for the long-term good of the country. Whether as a direct result of the book or not, replanting began, and the trees were reaching maturity almost a century later, just as the Seven Years War began. Neverthless, the Navy Board anticipated shortages and made discreet enquiries about what supplies were available, hoping not to alert the timber merchants and thus drive prices up. Inevitably, they were unsuccessful. 'The persons we employed for this purpose had scarcely gone half through the first two counties in which they were to begin their enquiry before they were interrupted by the proprieters of estates of timber, and we were obliged to drop our intention of proceeding further.'[3]

Not all timber was suitable for naval construction. Oak was much preferred for building the hull though a few frigates were built of fir later in the century, but they soon decayed. After 1800 the teak forests of India were exploited, but at the time of the Seven Years War they were still untouched. Furthermore, shipbuilders had a strong prejudice in favour of English oak for the frame timbers, though they might allow foreign oak for the planking. In 1760 only 345 loads of ordinary oak timber were imported to Britain, compared with 4,105 loads of plank from the Baltic alone. English oak was regarded with almost religious veneration, and it is no coincidence that the naval anthem *Hearts of Oak* was published in 1759, the same year that the *Valiant* was launched. In the 1800s the navy was to be forced to import much larger quantities of timber from the Baltic, and was to find that it was perfectly suitable for shipbuilding; but in the middle of the eighteenth century this was not considered.

Even among English oaks, the shipwright was selective. Straight timber had its uses, for sternposts, deck beams and other parts; but the most valuable pieces were the curved 'futtocks' which made up the frames of the hull, and the L-shaped 'knees' which braced the deck beams against the sides. Such timber had to be grown naturally, and the grain had to follow the shape of the futtock or knee. This wood, known as 'compass timber', could fetch high prices at market, but it was difficult to grow. Intensive cultivation was of little use because forest oaks tended to grow straight up, to compete for sunlight. The best oaks were those growing in hedgerows or in isolated sites. Furthermore, a large tree took 80 years or more to reach maturity. There was always the temptation for the landowner to cut it down before then and sell it for house building, or for small ships but the traditions of inheritance of the time tended to keep the great landed estates intact, and encourage landowners to plant for their grandchildren or great-grandchildren; this did much to help the long-term timber supply.

A ship of the line as big as the *Valiant* required a vast quantity of timber. No figures are available for the *Valiant* herself, but those for her sister ship the *Triumph* are known. That ship consumed 3,028 loads, of which 1,355 were of compass oak. Another 597 were of straight oak for the sternpost and

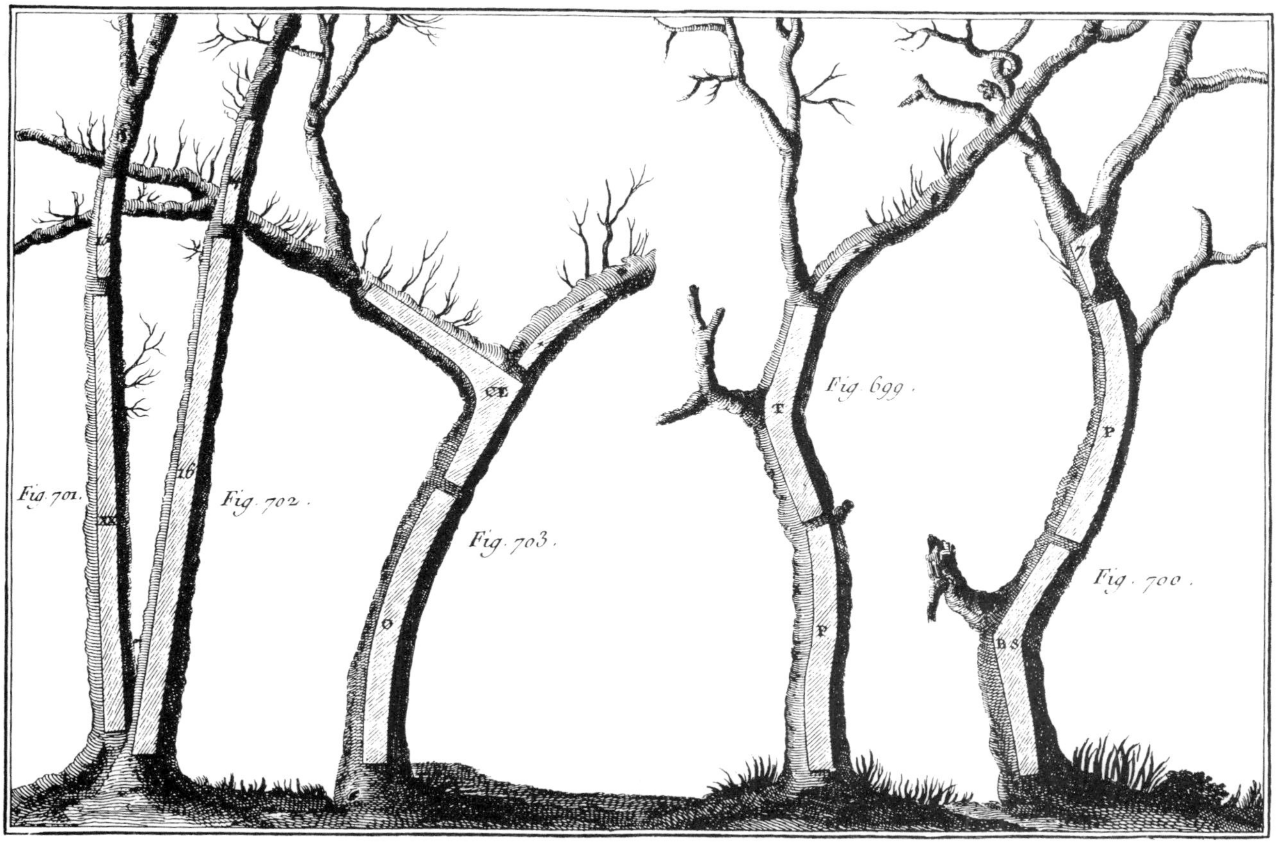

Selecting timber from trees.
(From the French Encyclopedie Methodique Marine*)*

other parts, and 62 loads of elm were used in the keel. One hundred and thirty-five loads of fir were consumed for sundry purposes, along with 61 loads for square knees, and 76 loads for raking knees. Most of the rest was oak plank of various thicknesses – 10in for the wales, 4in for the underwater hull and 2in for the decks. But the *Triumph* was surprisingly economical; smaller 74-gun ships, such as the *Mars, Thunderer* and *Canada* consumed 3,342, 3,713 and 3,405 loads respectively. In contrast the *Royal George*, a 100-gun ship, needed a total of 5,760 loads; and a 50-gun ship needed 2,100 to 2,450 loads.[4]

Sources of Timber

Some naval timber was supplied from the Royal Forests, but not nearly enough. The term 'forest' is slightly misleading, for it denoted an area set aside, many centuries ago, for the recreation of the King and his courtiers, particularly for hunting. Special laws applied to the inhabitants, and they were not allowed to do anything which might interfere with the game. Such an area was not necessarily completely wooded, though some were. In fact, only two of the Royal Forests, the New Forest and the Forest of Dean, were still supplying substantial quantities of naval timber by the second half of the century. By 1771 the New Forest was supplying about 870 loads to Portsmouth every year, out of an annual peacetime consumption of 6,000 loads; while the Forest of Dean was supplying about 600 loads to Plymouth out of 5,000 consumed annually at that yard.[5]

As a result, the navy turned to private landowners, usually dealing with them through timber merchants. There were still substantial woods in the Weald of Kent and Sussex, and these were close enough to supply the Chatham yard. However, there was competition from the iron founders of the area who used charcoal for smelting, and from those who procured timber for house building in London. Some oak could be found in Hampshire and in Essex and some came from further north. In 1757 it was reported that Messrs Richardson and Watson, under contract for supplying Chatham, had three vessels at Hull, with 'near 300 loads of large

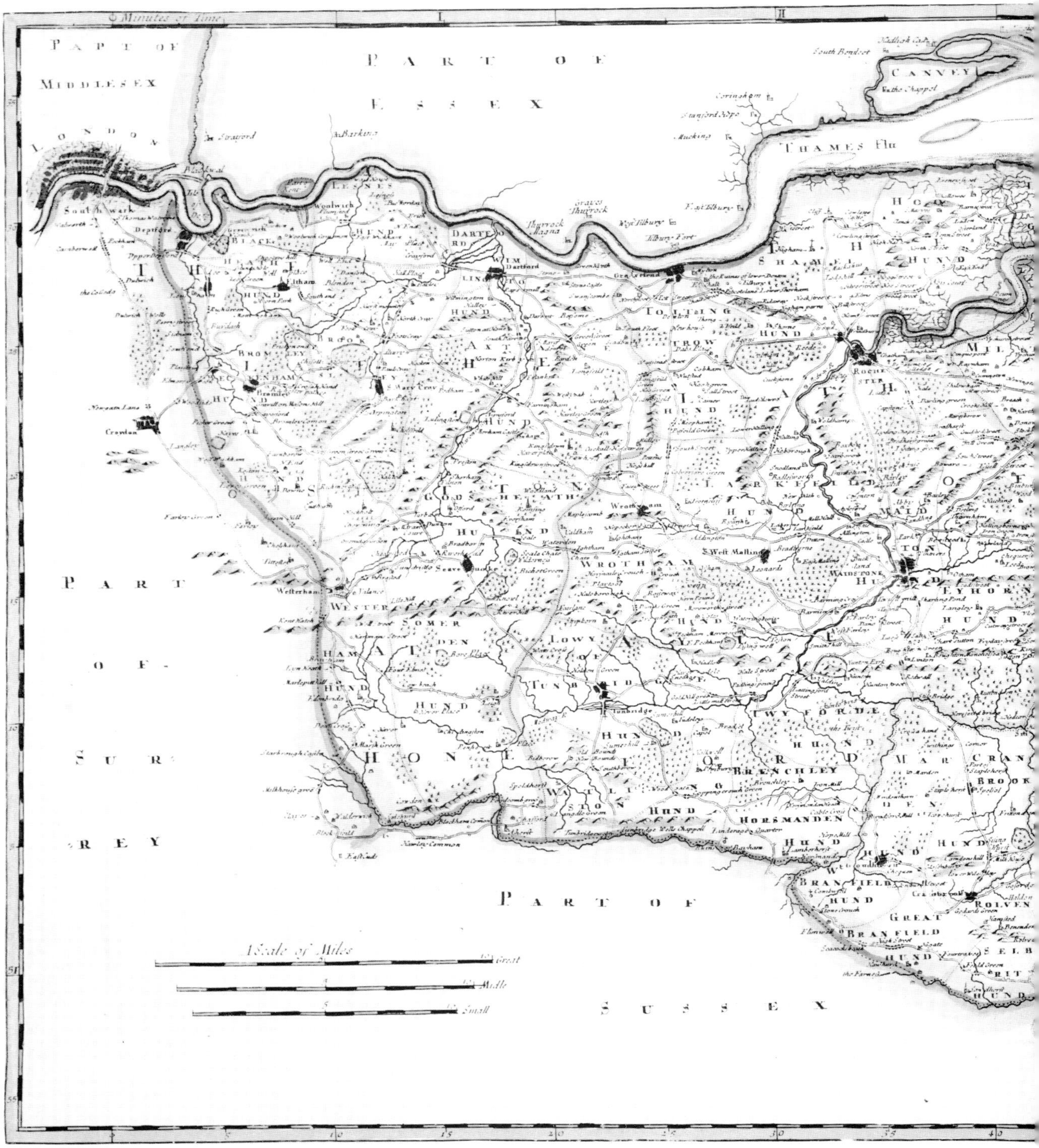
Minutes of Time
PART OF MIDDLESEX
PART OF ESSEX
LONDON
Southwark
Deptford
Woolwich
Dartford
Gravesend
Tilbury Fort
CANVEY
Rochester
Maidstone
West Malling
Wrotham
Tunbridge
Westerham
Croydon
Bromley
Eltham
HORSMANDEN
BRENCHLEY
PART OF SURREY
PART OF SUSSEX
A Scale of Miles
Great
Midle
Small

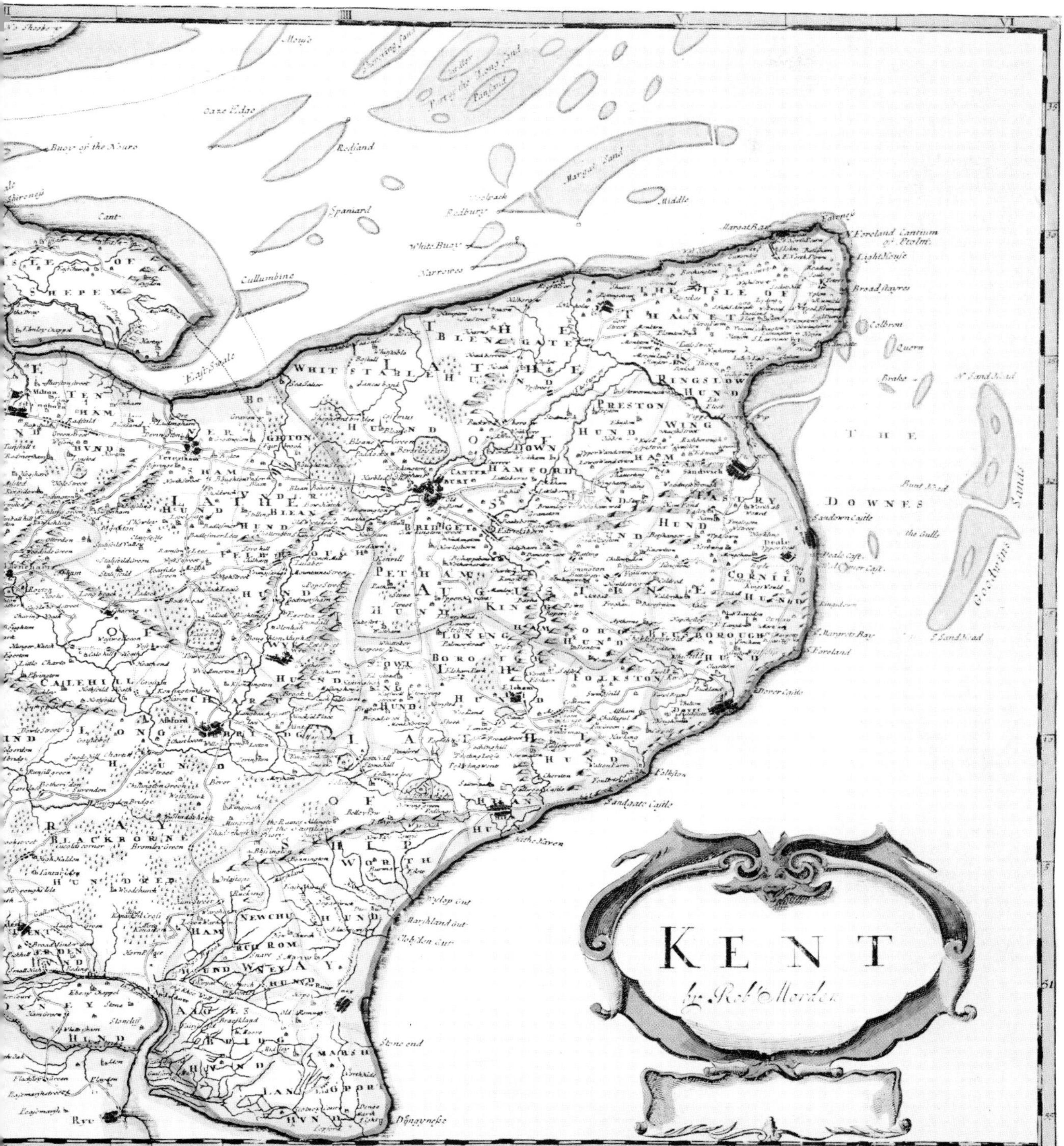

Kent in the early eighteenth century, showing the wooded areas.

Felling trees for shipbuilding in the New Forest in 1798. (British Library)

Timber wagons.
(From Pyne's Microcosm, *1803)*

timber and knees', and they requested a naval escort to protect them from French privateers in the area.[6]

Contracts

Timber was bought by the Navy Office in London, and contracts were placed for a period, usually of one year, specifying delivery to one or other of the dockyards. Contractors were invited to tender, and the contracts went to the lowest bidders, provided they were considered reliable. The system of government finance meant that money was always paid some time after delivery, so the navy was permanently in debt. However, the credit of the government was well established and there was no shortage of contractors. Wartime scarcity, on the other hand, soon caused the prices of timber to increase. Oak plank rose only slightly, from £6.0.0 per load in 1738 to £6.12.0 (£6.60p) in 1760. Straight oak timber rose much more steeply, from £3.0.0 in 1738 to £4.5.0 (£4.25p) in 1760.[7] The timber contractors were often accused of corruption and of forming rings to keep up the price of timber; though this was evidently less of a problem during the Seven Years War than in the 1800s. Ideally, the Navy Board would have liked to keep up several years advance supply in the yards, allowing time for seasoning, but the principle of keeping a three years' supply was not finally established until 1771, and during the Seven Years War ships were built very quickly, so that stocks were often depleted.

Government purveyors went into the Royal Forests and marked the trees they considered suitable with the 'broad arrow' – the simple arrowhead which had been the mark of government property for many centuries. Tree fellers usually worked in pairs, with axes swinging them alternately, to a regular rhythm. Those of the Weald of Kent had a particular style of axe, with cheeks on the head, and a relatively short wooden handle.

Imported Timber

There were two major sources of imported timber: Baltic masts and plank and North American masts. 'East Country' oak plank had been brought in since the Thirty Ships programme of 1677, and came via Danzig [now Gdansk], Memel and Riga. Fir masts also came from these ports, as well as Stettin and St Petersburg. The need to ensure timber supplies from the Baltic, as well as supplies of other goods such as tar, had a profound effect on British foreign policy. Earlier in the century, Britain had become suspicious of the rising power of Sweden and had sent fleets to the area on several occasions in order to maintain the balance of power and keep open the trade routes. In the intervening decades, Britain considered very carefully her relations with the Baltic powers of Sweden, Denmark, Prussia, Poland and Russia.

The British trade in the Baltic was conducted by agents known as factors, often living together in the same district of the port, and protected by a consul. The timber could be bought very cheaply from the forests of the region but considerable costs were incurred by its transport. To make this cheaper, the timber was mostly cut to shape before loading, either in the form of plank, or as the cores, or 'spindles' of masts of specific types.

The vast forests of the British colonies in North America were still unexploited, but the Navy Board was reluctant to use American timber in the hulls of its ships because New England oak was considered to be of poor quality. However, pine for masts was of considerable importance, and made up a large proportion of the navy's stock. The British Parliament passed numerous acts to protect the forests in its American colonies, some of which were greatly resented by the colonists. More positively, bounties were paid to the suppliers of naval stores such as tar, turpentine, hemp and timber. The acts were enforced by the Surveyor General of His Majesty's Woods and Forests in America – a post which was held by the Wentworth family, after a certain amount of bribery, from 1743 onwards. Suitable trees were marked with the broad arrow, and could be exported only to Britain, where the navy had first claim on them. The supply of North American masts largely ceased after the American colonists began their revolt in 1775, and after that the navy turned increasingly to the Baltic for supplies.

A model showing a crane and a timber hoy at Deptford Dockyard in the 1770s. (The Science Museum)

Transport

Before the age of steam railways, land transport was generally to be avoided, especially for heavy materials like naval timber. In Britain the main roads were gradually being improved by means of the turnpike system, but few good roads led into the forest areas. It was necessary to get timber to a navigable waterway and this was not easy; and it was commonly assumed that any timber more than 40 miles away from water transport was useless for the navy, as the costs of moving it were prohibitive. In the first stages of its journey to the dockyards and shipyards, a tree might simply be hauled along the soft ground by a large team of horses. Specially designed timber wagons were also used. Some were simply two-wheeled vehicles, with a yoke between, and a large tree could be slung underneath and half-dragged along the ground. Others were four-wheeled, and carried the tree on top. During earlier periods, for example during the 1677 programme,

A model of Chatham Dockyard in the 1770s, showing the space between the officers' terrace and the storehouses taken up with stacks of timber. (Author's photograph, courtesy of the National Maritime Museum)

shipwrights had been sent into the forests to cut the frame timbers of the ships to their finished size. However, this was an uneconomical use of labour and died out in the eighteenth century. Straight timber, such as materials for planks, could be cut to a square section before transport, but compass timber was usually left unfinished, with the branches lopped off.

Water transport was the most practicable, particularly as some rivers were being made navigable for the first time and canals were being dug. Some Hampshire timber was floated down the River Wey from Guildford, to join the Thames at Weybridge. From thence it could be sent to the yards at Deptford, Woolwich, Sheerness and Chatham. In Kent, the River Medway had been made navigable as far up as Tonbridge by 1740 and timber could be brought to Chatham by that route.

Logs could simply be floated downstream, and this tended to happen in the earlier stages of the journey. When it reached the main rivers, timber was more likely to be put into barges, mostly of local design and well adapted to local conditions. On the upper Thames, for example, a vessel of very angular construction was used which carried a single squaresail for use down wind, but which could be towed by horses or men, and even rowed when conditions demanded.

Timber seasoning sheds in Chatham Dockyard.

When timber was transported by sea, either along the British coast or from America and the Baltic, it was loaded into ships which, under the terms of the Navigation Acts, had to belong either to Britain or to the exporting country. Some of these vessels were specially designed. Baltic timber ships were of 250 to 300 tons, and able to carry about 300 loads of timber. They were usually of shallow draught to allow them to enter the Baltic harbours. English ships were less specialised, and were often veterans retired from other trades such as West Indian sugar or coal from northeast England. In wartime they travelled in convoys escorted by naval sloops and frigates. The Baltic convoys were very large, and in the 1800s they sometimes consisted of several hundred ships, escorted by two or three warships.

Inspection and Storage

On arrival at the dockyard, timber had to be inspected by the officers of the yard. A minor official known as a 'timber taster' inspected each piece, looking for knot holes, shakes (cracks) and other defects. It was measured and marked with his initials, and then put into store. In 1801 the office of 'Timber Master' was created at Assistant Master Shipwright level, in an attempt to make the system more efficient.

Whether timber was brought by barge or ship, it was unloaded by cranes along the waterfront. These were manually powered and came in several types; some being operated by a capstan, others by a treadmill. On some cranes, known as swinging cranes, the whole structure was pivotted with a jib that was an integral part of the body. Others had the treadmill or capstan enclosed in a small shed with a light moveable jib in front. The swinging crane seems to have been most common by the middle of the eighteenth century and certainly these were in use at Chatham. Having been unloaded, the timber was then moved around the yard by teams of horses.

In ideal circumstances timber would be stored in the yard for some time, partly in order to season it, but partly to keep up a supply for future use. During the Seven Years War this was impossible because ships were needed urgently for the fleet and timber was used almost as soon as it was acquired. Indeed, many ships had to be built with 'green' timber and their lifespan was inevitably short. An extreme case was the *Ardent* of 64 guns, completed in 1764 at a cost of £23,000, and found in 1771 to be 'in a total decay, her timber and planks rotted almost irrecoverably'.[8] The Earl of Sandwich commented on the shipbuilding effort of the war, 'Though we built as fast as possible in the King's yards, and employed all the merchant yards that could build a ship for us, it is to be observed that we finished with exactly the same number with which we set out'.[9] To be fair, however, the new ships built during the war were much larger and better designed than the ones which they replaced.

In 1771 it was decided that much better care should be taken with timber and that a proper programme of seasoning be introduced. Special timber seasoning sheds were to be built at all the yards to a standard design and the last survivors of these are still to be seen at Chatham.

Before the 1771 reforms timber was normally stored in the open and large spaces in the yards were available for this. It was stacked in square piles, often 15ft or 20ft high and the different kinds of timber were separated; some was uncut timber, and some had been cut on two sides only ('sided'). Compass timber was stacked in different piles from straight timber and, of course, different kinds of wood were stacked separately.

6

Cutting and Fastening Timber

The ship of the line was perhaps the most sophisticated single construction which at that time had yet been devised by man, but the tools and techniques used to build it were mostly simple and traditional. Tools were largely unaffected by the new metal-working processes of the industrial revolution and the shipwright was still a true craftsman, not a part of a production line. The shipwright, and his colleague the sawyer, used purely manual labour to fashion the frames, beams, planks and knees of the hull, after which the shipwrights put the pieces of the hull into place, and fixed them there by means of wooden trenails and metal bolts.

The Sawyers

The sawyers were the first to work on the timber, and it often came to them in its prime condition with the bark still on. If necessary, they might trim the top and bottom flat with an adze to give a stable base and a flat working surface. Heavy sawing work was usually done over a pit in the ground, rather than having the log raised above the ground by supports, as was done in most European countries at this time. Smaller logs were placed across the pit to support the one being cut which was kept upright by wedges and further held in place by means of 'dogs' – large staples

Sawyers at work. The hands of the bottom man can just be seen. (From The Book of English Trades, *seventh edition, 1818)*

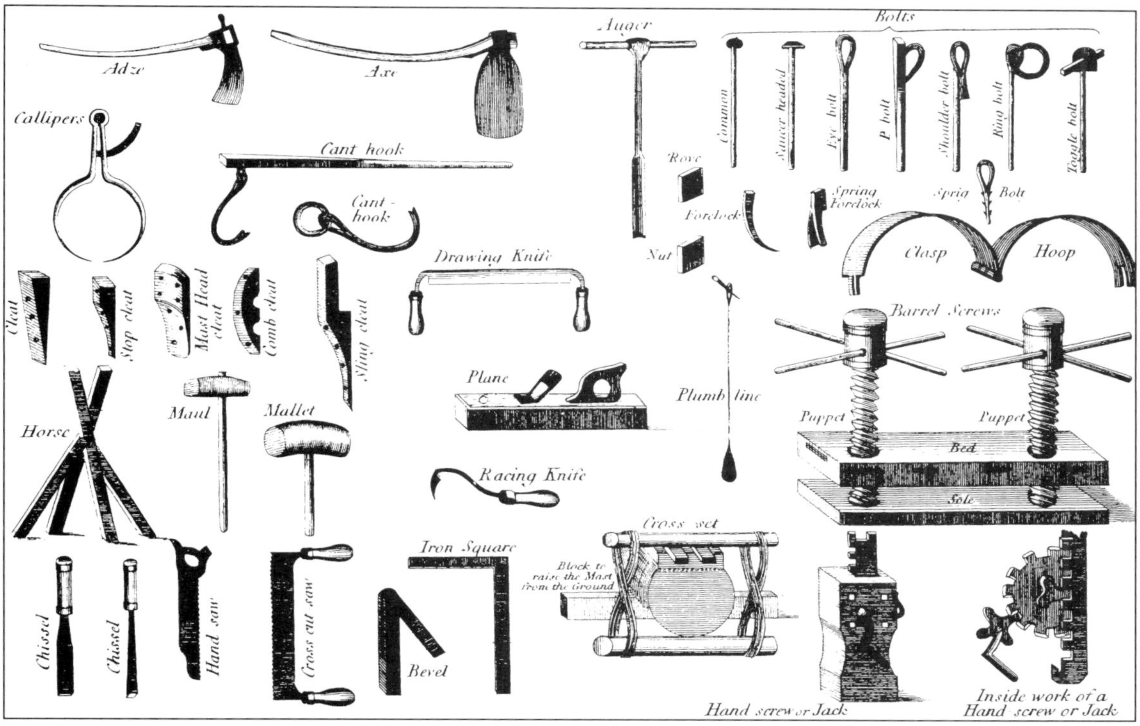

A selection of shipwright's tools. (From Steel's Rigging and Seamanship of 1795*)*

which were fixed into the side of the log and the top of the woodwork of the pit.

Sawyers worked in pairs, one standing on top of the log and the other beneath in the pit. In the dockyards the top sawyer was better paid as he was regarded as the more skilled of the two. The pit saw was 6ft to 8ft long and either had a narrow blade enclosed in a frame to tension it, as used for cutting curves, or a simple deep blade for straight cutting. The top handle on a saw was known as the tiller, and extended about 2ft above the blade, curving backwards. The lower handle, known as the box, was a kind of peg which passed through the blade.

The Work of the Shipwrights

The sawyers' work was largely to 'side' timber – to cut flat surfaces on two opposing sides. This done, the wood was then available for the shipwright to 'mould' it – to cut it on the other two sides, often in a curved shape according to the patterns received from the mould loft. The shipwright had a more varied tool kit than the sawyer. It included an axe, a set of augers for drilling, a draw knife, chisels, gouges, hammers of different kinds, a maul, a brad and perhaps caulking irons, scrieve hooks for work in the mould loft, and cant hooks for lifting timber. He had a 'small saw' which he may have used for the moulding of timber and he also had his adze, the most characteristic tool of his trade. This was similar to an axe except that the blade was set at right angles to the handle, not in line with it. The blade was slightly curved and a 'peg poll' behind the handle was used for driving nails into the hull. Its wooden handle often had a double curve which allowed the instrument to be very precisely balanced. The adze could be used for several purposes: for cutting the hull timbers to shape, especially those near the bow and stern which needed to be bevelled; for smoothing out the run of the assembled timbers, to make them fairer; and to trim the planks of the hull and decks, and to plane their surfaces. The skill of the shipwright with his adze was almost legendary. 'In coarse preparatory works, the workman directs his adze through the space between his two feet, he thus surprises us by the quantity of wood removed; in fine works, he frequently places his toes

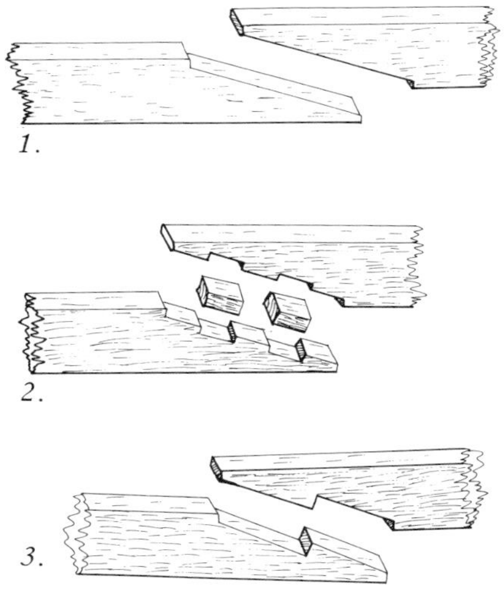

Types of Scarph.

1. Plain scarph

2. Coaked scarph

3. Hook and butt

over the spot to be wrought, and the adze penetrates two or three inches beneath the sole of his shoe, and he thus surprises us by the apparent danger yet perfect working of the instrument.'[1] For surfaces which were not horizontal, such as the sides of a ship, he held the end of the handle on his knee, and used that as a pivot.

Scarphs

Apart from his skill in cutting timber, the shipwright required numerous ways of fitting pieces together. He called this 'faying'; to fay was 'to fit any two pieces of wood so as to join close together'.[2] Few parts of a ship could be made from a single piece; frame timbers, keels, false keels and keelsons, runs of planking and deck beams all had to be constructed out of several pieces of timber. Planks did not usually need to bear much stress in relation to each other, so it was normal to lay them end to end without any direct connection between them. Until the second quarter of the eighteenth century it seems that the futtocks which formed the frame timbers were treated in a similar way; there was no direct join between them, and they were held in place in relation to one another by ribbands during construction and by the planking thereafter. By the 1750s, however, it was common to cut out a triangle from the inner end of each of the futtocks to be joined, and to place a wedge, or 'chock' across the join. By about 1780 the two corners of the triangle were cut off the chock, so that it became five-sided. Bolts were then placed through the chock and the timbers; by a contract of 1779 each join was to have three bolts of 1¼in diameter on each side.

In the case of relatively straight timbers, such as deck beams and the keel, overlapping joints known as scarphs were used. In a plain scarph, the ends of the two pieces to be joined were simply cut at an angle, perhaps with the end of the piece trimmed off, and a step left in the corresponding position on the other piece, to prevent sliding. In a slightly more sophisticated form, the scarph was 'hooked' – one or more steps were placed along the length of the join to add further longitudinal strength, and this form of scarph might have wedges driven in from each side at the step, in order to make it tighter. A scarph might also be 'tabled' whereby a rectangular or mortice was cut out of one of the faces and matched with a tenon in the other face. In another variation, holes would be cut out on both faces, and fillets known as coaks fitted. Either way, the tabled scarph had the same advantage as the hooked scarph, with the further one that it also helped brace the join against sideways movement.

Tenons and Dovetails

Where two pieces of almost equal importance met at approximately right angles (such as the sternpost and the keel) one was tenoned into the other. A projected tongue was left on the lower end of the sternpost, and a mortice was cut in the top surface of the end of the keel. Where the pieces to be joined were in the horizontal plane, as with the

A horseshoe plate on a model of the 1790s.
(Author's photograph, courtesy of the National Maritime Museum)

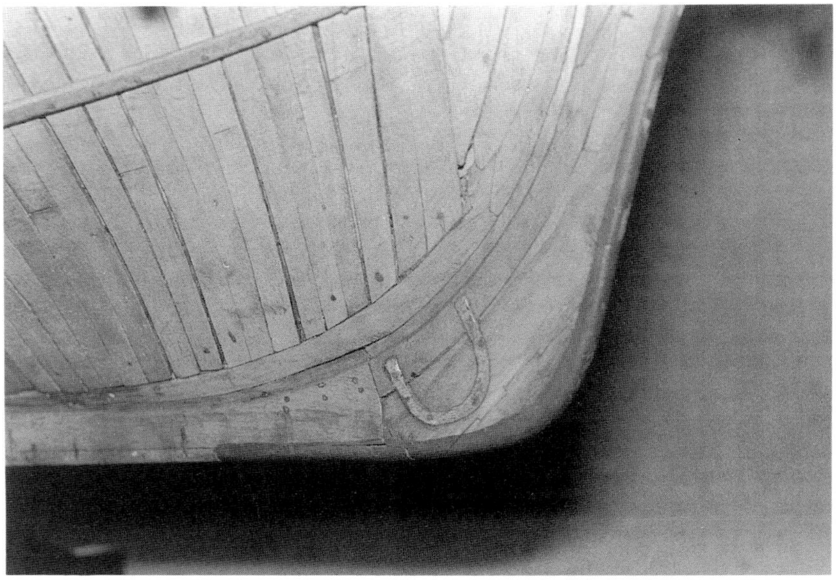

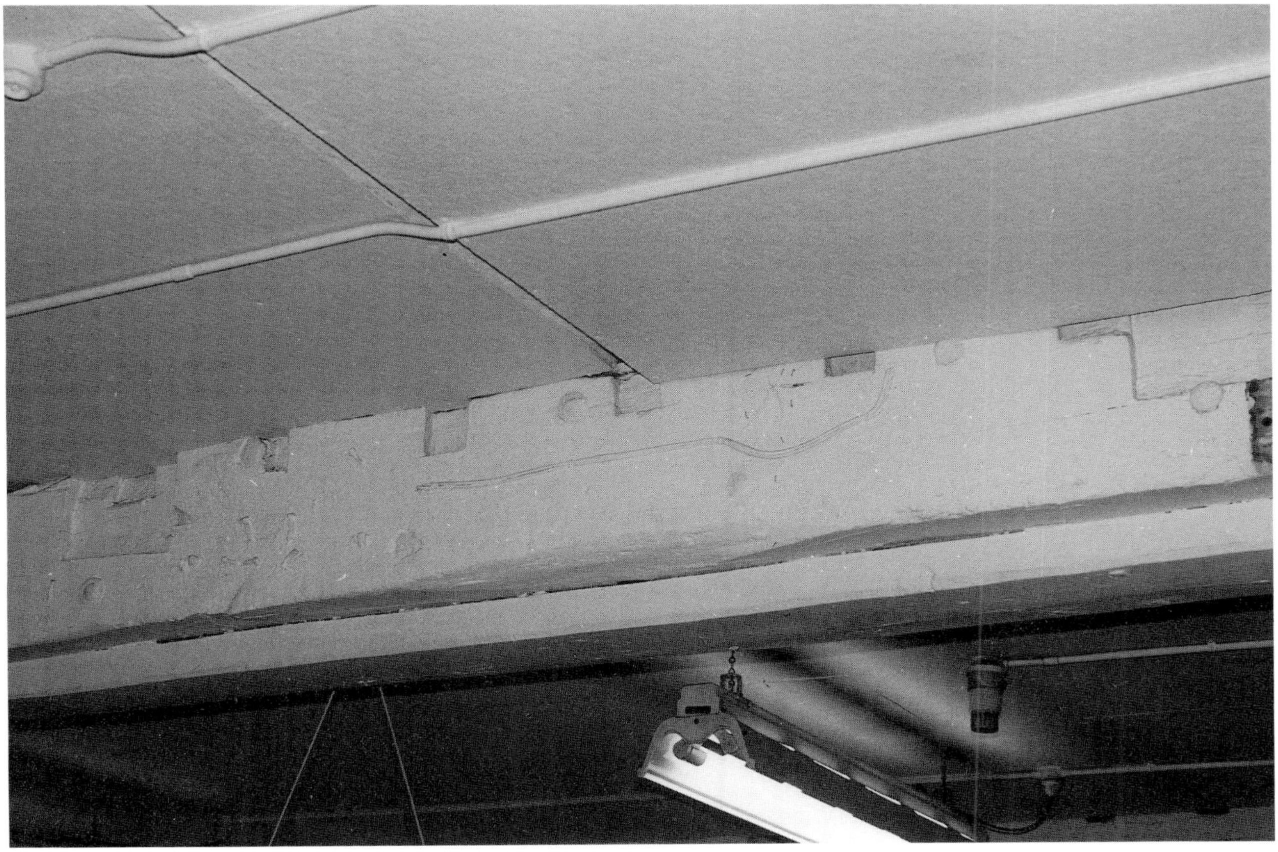

Remains of tenons in old deck beams in the sail loft at Chatham.

beams, carlines and ledges which supported the decks, a dovetail joint might be used. The tongue which projected from the end of the smaller piece, such as a carline being fitted into a deck beam, was angled outwards towards its extremity so that it would not be pulled out of position too easily, and a carline could thus serve to hold two deck beams the same distance apart against the stresses of the hull. However, dovetails were not always used, a simple tenon join being employed instead.

On a few rare occasions, the mid-eighteenth-century shipwright used iron to help bind pieces of timber together (other than in the form of nails and bolts). The most notable places were at the bow and stern, near where the stem- and sternpost joined the keel. The sternpost, rising almost vertically from the end of the keel, was held in place by the 'dovetail plate', a flat piece of iron shaped like two dovetails joined together, with a narrow neck covering the line of the join. One was fitted on each side, and bolts were passed through the post to hold them in place. At the bow the 'horseshoe plate' was used to hold together several parts of the stempost and the knee of the head. Its name describes the shape quite well and, as with the dovetail plate, one was fitted on each side. Both types of plate were made from copper alloy after coppering became common in the 1780s.

Trenails

The shipwright employed three main methods to fix the various timbers in place. Nails (usually known as spikes) were not trusted for the most vulnerable and important parts of the structure, but were used for deck planks and for the ribbands which held the timbers in position until the planking was done. Bolts were used for most of the internal structure, including the frames, knees and riders in the hold. Wooden pegs known as treenails (generally spelt trenails and pronounced trunnels), held most of the rest of the structure together, including the planking of the hull.

Trenails had several advantages over metal bolts and nails. When used to hold on the planking of a ship, they offered no hindrance to the shipwright's adze and, unlike iron, they did not rust. They were made from oak, and according to one contemporary authority, 'it is necessary that the oak of which

Trenails from the Invincible. *One is unused and still has its eight-sided shape. The other has been rounded and has a wedge in the end.*

they are formed should be solid, close and replete with gum to prevent them from breaking and rotting in the ship's frame'.[3] The timber of the trenail needed to be well dried so that it would swell when exposed to moisture and make a very tight fit. Trenails were 1in in diameter for every 100ft of the ship's length, so those used in the *Valiant* would have been 1½in.

In the first instance trenails were made by 'trenail mooters', generally aged shipwrights who were unable to do the heavier work. Each was cut to an eight- or sixteen-sided section and issued to the shipwrights and only when it was to be used was it trimmed to a rounded shape and banged in to the appropriate hole. Sometimes trenails were simply left like that, without further fixing, but there is evidence that the trenails of the 1750s each had a wedge hammered into a groove across the diameter to secure them further.

Metal Fastenings

Until the 1780s iron bolts were used in many parts of the frame. They secured the scarphs of the keel, false keel and keelson. Long bolts were needed to hold the different parts of the knee of the head (see page 00) together, as they passed through the whole depth of the various timbers. Shorter bolts were used for the chocks of the frame timbers, and for the scarphs of the deck beams. The knees were also held in place by bolts. One group passed through the depth of vertical arm of the knee to meet the frame timbers, and another went through the thickness of the horizontal arm to join the deck beam which was laid alongside it. Bolts also secured the thick riders and breast hooks which were fitted inside the hull for additional stiffening.

Some bolts, known as 'rag bolts', were more like nails or spikes in that they did not pass all the way through the pieces of timber to be joined but had barbs along their sides to secure them. More common, though, were the bolts which passed all the way through the timbers. Each of these had a round head, and was cut to length according to the size of the pieces to be joined. It was passed through the hole from the outside and a round washer, the 'rove', was passed over the end; then an iron wedge, the 'forelock', was placed through a hole in the end of the bolt to lock it in place.

From the 1780s, when copper plates were fitted to the bottoms of ships to protect them against weed and worm, it was soon found that electrolysis between the copper and the iron caused rapid decay. Attempts were made to separate the copper from the iron by means of tar and paper laid between the copper and the planking but suc-

A bolt with washer and forelock, used in the constuction of the sail loft at Chatham.

cess was limited and short-term and eventually it was decided to eliminate iron bolts from the underwater hull. Pure copper was unsuitable for bolting as it was too soft but after many experiments a suitable copper alloy was found. During a programme which began in 1783 the old iron bolts were driven out of every ship in the navy in turn and replaced by copper alloys.

Although bolts were favoured for the main parts of the structure various kinds of nails were used for particular purposes in the construction of the ship, each type being designed for a special job. Spike nails held the deck plank in position; sheathing nails held on the sheathing boards, as fitted over the planking of ships sent to tropical waters; wolding nails held on the ropes which were bound about the masts; scupper nails held the leather scuppers to the ships sides; and the boatbuilder who made the ship's boats had many kinds of nail of his own.

Staples were used to join the false keel to the main keel, being hammered in from the sides of the parts to be joined. This join was deliberately made quite weak so that the false keel could break away easily if the ship ran aground, without causing damage to the main structure; and, among other purposes, staples were also used to hold the parts of the gunport lids together.

A shipwright boring holes with an auger.
(From R Dodd, Days at the Factories, *1843)*

Boring Holes

Staples and nails could sometimes be hammered into place without any preparation though large nails did usually require a hole to be drilled at least part of the way. Trenails and bolts obviously required holes to be drilled to their full length and to the exact diameter and these holes were drilled by means of augers. The modern twist drill was not known until 1770, and it did not catch on even then. It was reinvented several times, and did not become at all common until well into the nineteenth century. Instead, the shipwright used the 'pod auger', shaped rather like a hollow cylinder with one half missing, and with a cutting edge at its lower end, which might take various forms. For cutting relatively small holes he fitted his auger to a brace or 'wimble', similar to the modern brace in shape, but usually made of wood. Alternatively, he simply fitted a handle through the head of the auger, at right angles to it and by turning on the handle he could bore large and accurate holes.

As well as the main hole which fitted the size of the bolt used, it was often necessary to drill larger holes to countersink the heads and forelocks of bolts; this was especially true when the surface holding the bolt was to be planked over, as in the case of the bolts through the chocks of the frame.

The distribution of trenails and bolts about the structure of the ship might have appeared random to the casual onlooker, but it was, in fact, very deliberate. Care was taken that fastening holes were never placed in a single line. If all the trenails fixing a single plank were placed carefully in line, for example, then there was a risk that they would follow a line of grain and split the plank. Similarly in the vertical plane, too regular arrangement of the trenails might create a weakness in the frame timbers.

All these techniques of cutting, drilling and fastening had to be used on the hull of a ship of whatever size. The shipwright was expected to be familiar with all of them, except perhaps siding timber. In modern times, when 'skilled work' is largely broken down into separate parts and aided by machines, it is difficult to appreciate the range of work that a craftsman had to be ready to undertake.

7

The First Stage in Construction

Slips and Docks

Most ships were built on slips and launched from them into the water. Of course, the launching was itself a difficult operation and much skill was needed to prevent the ship being damaged by too quick an entry into the water, especially in a narrow river like the Medway at Chatham. As a consequence, some ships were built in dry docks rather than on slips, for the operations of floating out tended to be much slower and safer than that of launching. Building in dry docks was quite rare, as they were in short supply, especially in wartime, when they were needed for the cleaning of ships bottoms and for repairs. Certainly only the larger ships were built in docks. According to Falconer's Dictionary of 1769, 'Ships of the first rate are commonly constructed in dry-docks, and afterwards floated out'.[1] In practice it was never quite as simple as that. The *Britannia* of 1762, of 100 guns and 2,116 tons, was the largest ship ever launched from a slip at the time when Falconer was writing; and shipbuilders seem to have gained in confidence over the years, and built ever larger ships on slips. Even in the restricted waters at Chatham the *Royal George* of 100 guns and 2,286 tons was launched from no 1 slip in 1788. But in the 1750s, relatively small ships were still being built in the docks at Chatham and the *Valiant* was one of the few Third Rates ever to be built in a dock (apart from those rebuilt). However, it must be remembered that she was larger than any 90-gun ship built before 1755,

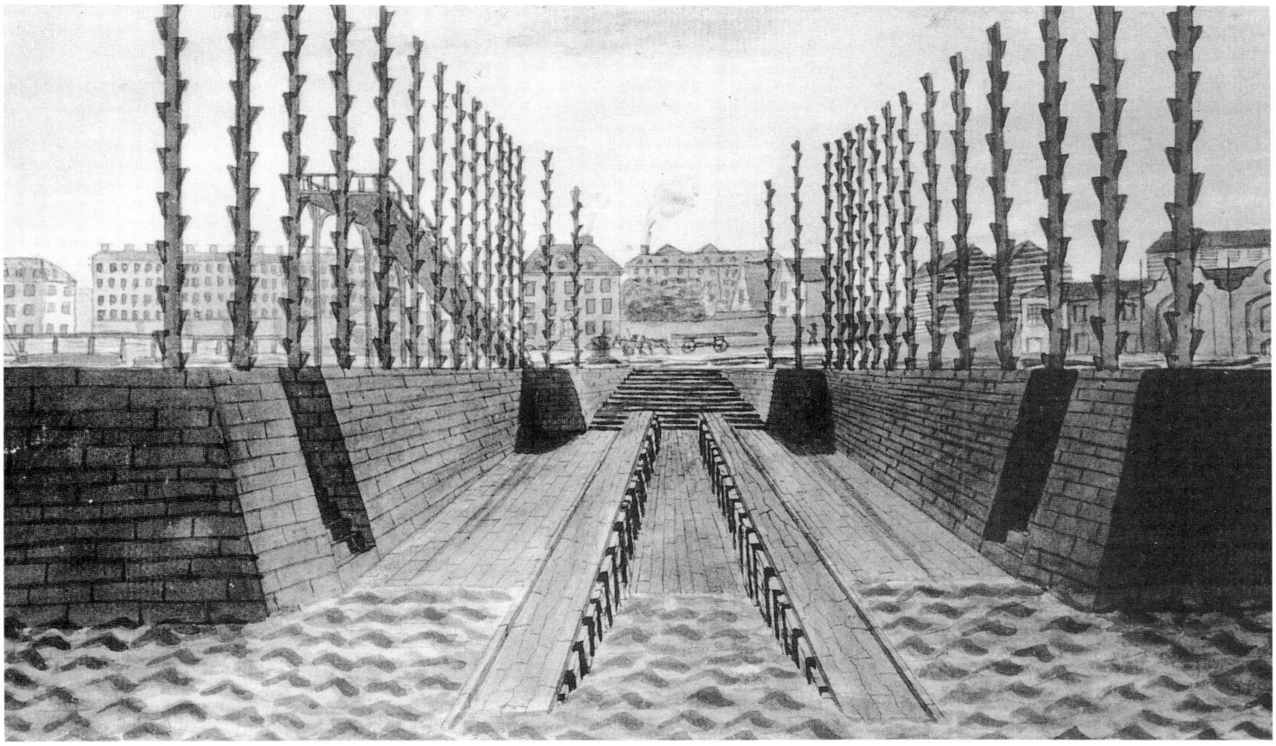

No 1 building slip at Chatham immediately after the launch of the Royal George *in 1788, looking towards the Commissioner's House.*
(British Library)

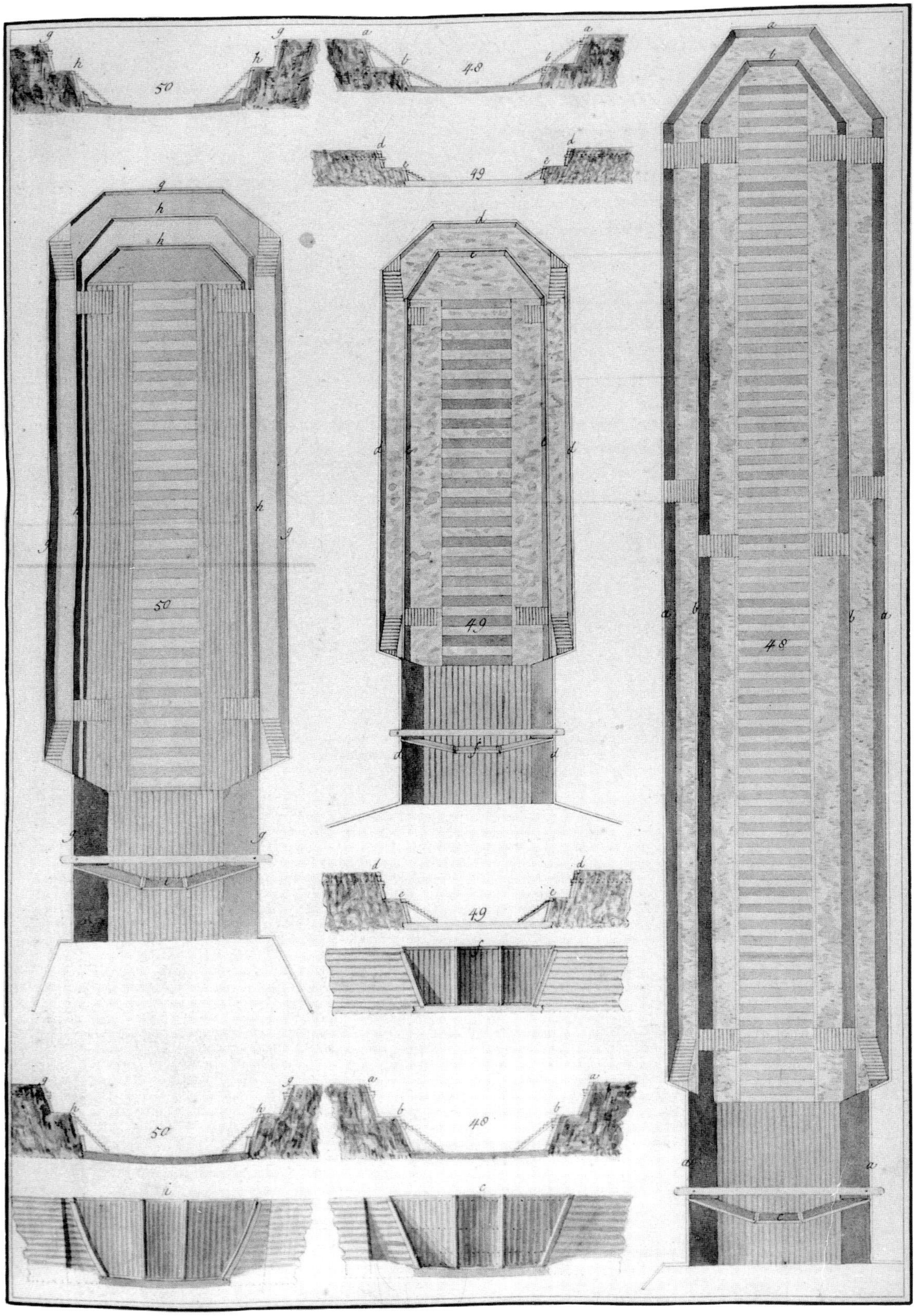
50
48
49
50
49
48
49
50
48

◄ *Dry docks at Chatham. At this stage most of them were still lined with wood, rather than stone. (British Library)*

and was only 3ft shorter than the *Royal Sovereign*, the largest First Rate afloat at the time.

A dry dock was essentially a hole in the river bank, with one end at the riverside which was fitted with lock gates. Tunnels, known as sluices, were dug under the dock to allow the water to drain out to the river at low tide, and to enter again at high tide when the ship was ready. Of course, docks had to be slightly larger than the ships that were to enter them and those at Chatham were about 200ft long, 60ft broad and about 25ft deep. Some docks were rounded at their inshore end, following the shape of a typical ship, though the Chatham docks were more angular. In cross section, a dry dock was much narrower at its base than at ground level, again reflecting the shape of a ship. By this time most docks had large steps along their sides which provided bases for the props which would support the hull. Docks built towards the end of the eighteenth century were lined with stone but those at Chatham were still made of elm (a timber which is most resistant to rot), reflecting the yard's declining status as a naval base.

A building slip was much shallower than a dry dock and a ship built there would tower above the ground. It was similar in plan to a dock, but its bottom was angled downwards towards the river, at about four degrees, to allow the ship to slide gently into the water on launch, and a small lock gate might be fitted at its end, to prevent water entering at spring tides.

A line of large wooden blocks was laid along the centre of the dock or slip on which the keel was laid. These had to be level because any irregularities in height would distort the shape of the keel which would be intensified as the heavy

A building slip, showing the keel blocks, vertical supports and platforms beside the slips. (Author's photograph, courtesy of the National Maritime Museum)

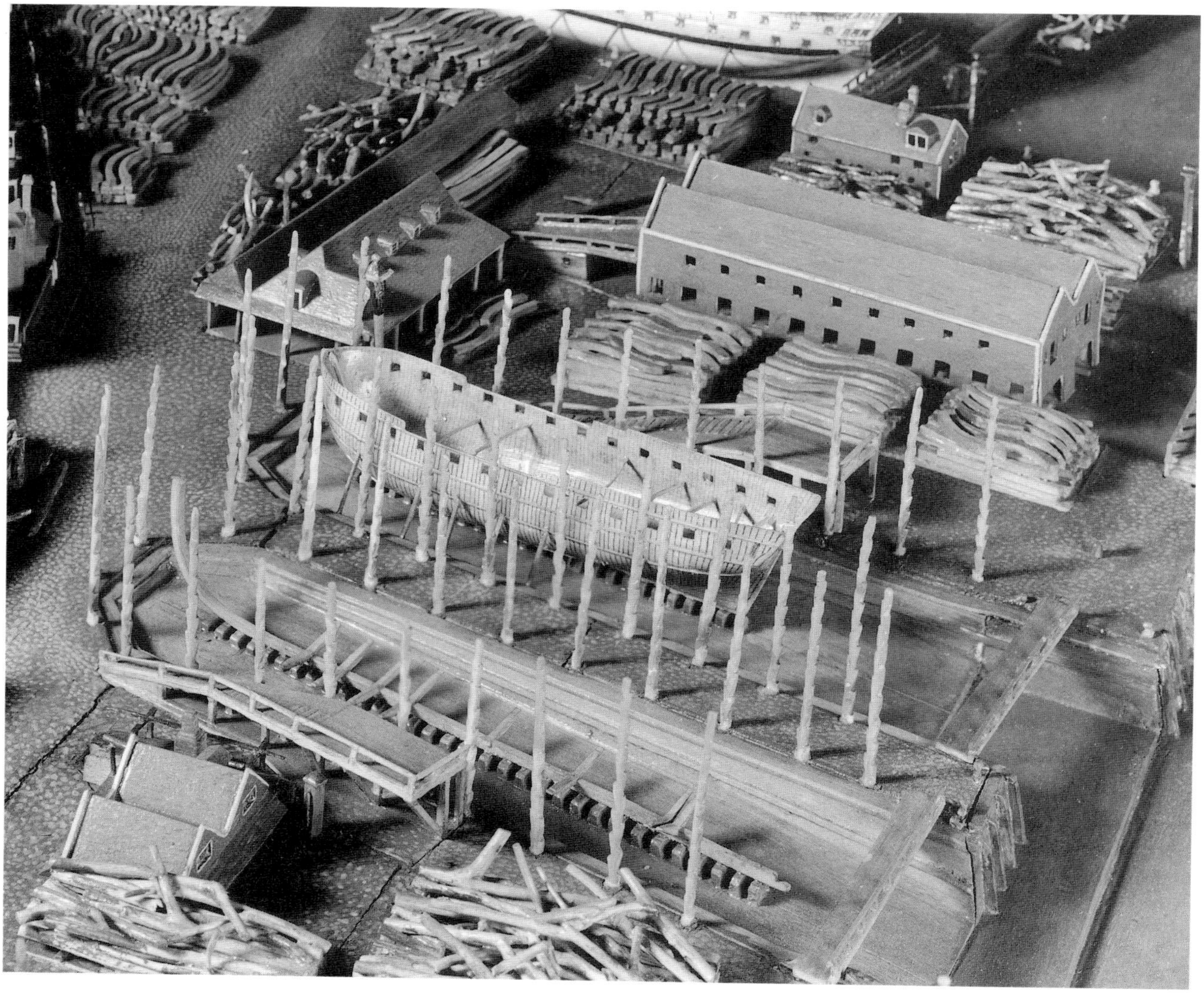

A slip at Woolwich, with the keel of a ship of the line laid and some of the floors crossed. In the background, a frigate is in frame. (The Science Museum)

timbers were placed on top and the keel ended up bearing most of the weight of the ship. Alongside each slip platforms were kept permanently erected, about 15ft high, and a ramp led up to them. They were used as a base for working, and for raising heavy loads. Capstans were sited round the docks and slips, partly for moving ships in and out of the dock, but also for lifting weights. Vertical posts were sited round the sides of each slip, with triangular wooden steps to form a basis for a system of scaffolding, and these were removed only when a ship was about to be launched. 'Sheers' were erected at each end of the dock or slips. They consisted of a pair of large straight timbers (probably old masts) lashed together at their head which, when fitted with a block and tackle, could be used for lifting. Often a sheer was set up at each end of the dock, with a rope running between them, and further ropes running forward and aft of the slip, rather like tent guys. The stem- and stern-posts could be lifted into place by tackles suspended directly under the sheers while the frame timbers could be raised by means of tackles from the rope running between them.

Laying the Keel

The *Valiant* was begun on 1 February 1758. This does not mean that she was 'laid down', but that the first parts of her keel were laid on the blocks. The laying of the keel had no great significance in the eighteenth century and does not appear to have been the occasion for any ceremony. Much more important, for accounting purposes

The parts of a stempost and stemson.
(From Fincham's Outline of Shipbuilding*)*

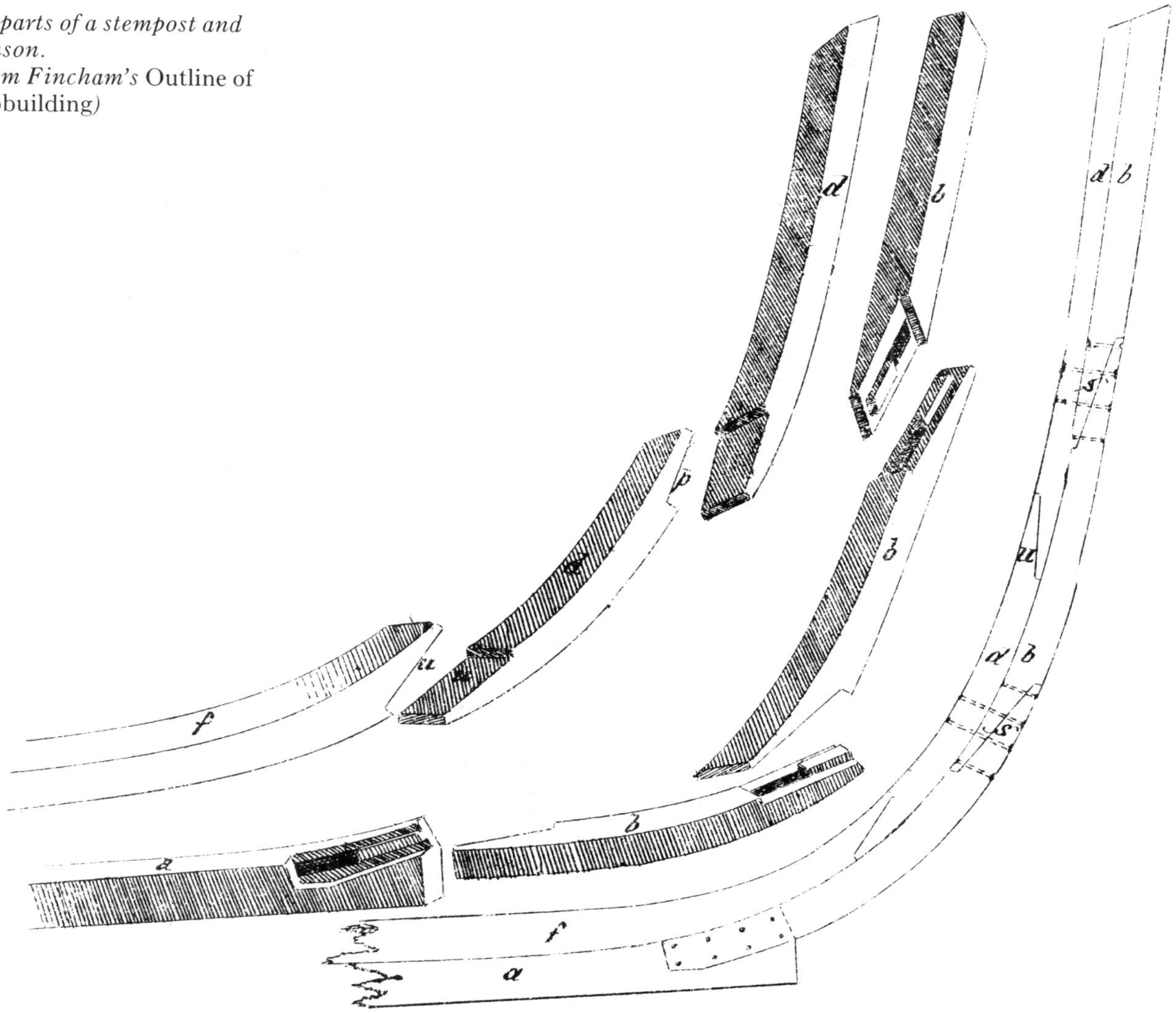

at least, was the date on which the first labour was employed on the ship, and clearly the pieces of the keel had to be made before they could be assembled in the dock. Few records are available for the exact sequence of events, but those from Portsmouth in the 1740s show that the term 'begun' was not synonymous with 'laid down'. For example, the Admiralty Progress Books record that the rebuilding of the *Tilbury* began on 8 February 1743. According to the 'Progress on the works at His Majesty's said yard the past week'[2] for 15 February, her rebuilding was taken in hand 'on the 8th inst'. However, the actual laying of the keel took place some weeks later, in the week ending 1 March; 'Bolted the scarphs of the keel. Fitted the upper false keel. Laid the keel on the blocks.' Therefore, it can be assumed that the laying of the keel of the *Valiant* took place a few weeks after 1 February, and that the dock was not needed until that time.

The earliest stage in the construction of the ship began with the keel, to which were added the stempost, sternpost, knee of the head and deadwood. All these timbers were in the same plane, vertical along the length of the ship, and of approximately equal thickness, so at this stage the ship was being built in one dimension, except that it was becoming common to add the transoms and fashion pieces to the sternpost before erection, and these provided the only three-dimensional element in this stage of the building.

The keel was long, straight and essentially square in cross section. It formed a backbone to which most of the ribs were attached, and it was the first part of the ship to be put in place. Unlike most parts of the hull it was made of elm rather than oak, as that tree tends to grow in very straight lengths and is less susceptible to rot. The keel of the

Deptford dockyard in the 1750s. Behind the ship which is ready for launching another ship can be seen under construction. The stern frame has been raised and the sheers for lifting it can be seen. Behind, a timber hoy is being unloaded and a team of horses is ready to take a large piece away. *(National Maritime Museum)*

74-gun ship of this period was 18in square at midships. The vertical dimension was kept constant throughout the length of the keel, but in the horizontal plane it tapered gradually, so that it was 14in wide where it met the stempost in the bow, and about 13in wide at the sternpost. In addition, the keel had a groove, known as the 'rabbet' (a corruption of rebate) just below its top edge on each side. This was intended to receive the edge of the lowest plank (the garboard strake), and it varied in shape along the length of the ship according to the angle at which the plank was intended to meet it. Amidships, the groove was rather like an equilateral triangle, with an apex pointing directly inwards. Forward of that, the angle of the garboard strake changed until it reached about 45 degrees to the horizontal, and the shape of the rabbet was altered to accommodate this. Aft of midships, the garboard strake became almost vertical, so that the rabbet became a rectangle removed from the top corner of the keel on each side.

The keel of the *Valiant* was 162ft long (the length of the keel 'for tonnage' was quoted in official figures as 139ft, but that was a purely theoretical figure used for calculation). No tree would yield a piece of timber that long, so the keel had to be made in several pieces scarphed together. Normally a keel of this size would be made in six pieces, of approximately equal length. The keel scarphs were vertical, with tables and coaks, and each scarph was 4ft 6in long, according to a contract of 1779. Each was to be bolted with eight bolts of ¼in diameter, and the surface of the join were covered with flannel or kersey, in order to protect it from shipworm.

The false keel was fitted under the main keel and would have already been laid. It was equal to the keel in width, but only 7in thick. It too had scarphs, in the horizontal plane, which were carefully arranged so as not to coincide with those of the main keel. The false keel gave extra depth to the ship, which helped prevent it from making leeway. It protected the main keel from damage if the ship went aground, and also helped in the defences against shipworm. The top of the false keel was heavily coated with a mixture of tar and hair, to prevent the worm penetrating the main keel.

The stern deadwood of a ship. This illustration dates from the 1820s, and the construction and shape of the rudder are different from the mid-eighteenth century.
(From Fincham's Outline of Shipbuilding*)*

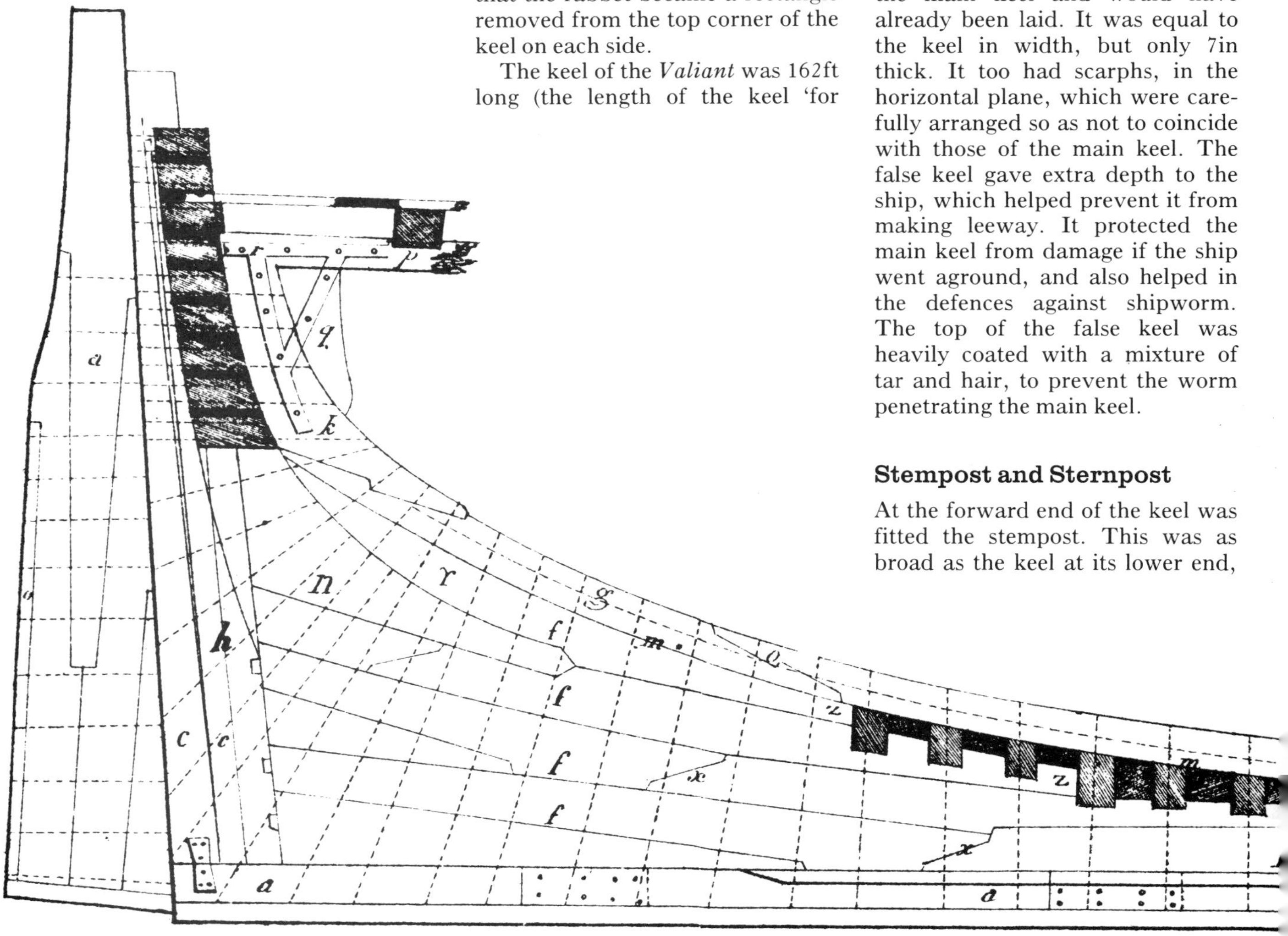

Stempost and Sternpost

At the forward end of the keel was fitted the stempost. This was as broad as the keel at its lower end, but it curved sharply upwards from it. It had rather less depth than the keel but it had a rabbet in its edges. The stempost usually consisted of about three pieces, scarphed together. It was fitted into the end of the keel by a rather complex join known as a 'boxing scarph', which was intended to prevent it from moving either up and down or from side to side. Immediately behind the stempost was another structure, of similar depth and thickness, known as the apron, or false stempost.

The sternpost was much simpler than the stempost, for it was a single straight piece of timber, raised at an angle of about 5 degrees from the vertical, and tenoned into the aftermost part of the keel. At its lower end it was as thick as the keel, but it widened after that and all the way up it had a rabbet for the ends of the planking. Until about half way up the post, the planking was almost parallel to the keel, so it was merely necessary to make the sternpost a little less wide forward of the after edge of the planking. Further up, the planking met the sternpost at an increasing angle, so the rabbet began to take on a triangular shape.

Inboard of the sternpost was the false sternpost. This was also a straight piece of timber, bolted to the forward side and tenoned into the keel. Its upper part was cut with grooves to receive the transoms. It helped strengthen the sternpost, and also formed part of the deadwood which supported the frames of the hull.

Transoms and Fashion Pieces

The term 'transom' presumably originated from the days of the early seventeenth century when even the largest ships had a square stern extending down to just below the waterline. The word still survives in modern boat and shipbuilding and refers to such a flat stern. In eighteenth-century Britain, however, the word had a further meaning. In large ships, transoms were straight horizontal pieces, placed one above another across the sternpost, and shaped to fair with the run of the hull timbers. The highest transom, the wing transom, was an important piece of timber because it had to support the weight of most of the stern galleries, counters and some of the timbers of the sides. It was the longest of the transoms, being about 30ft in the case of a 74-gun ship. Its after edge, which defined part of the shape of the stern, was cut with a slight convex curve, while its inner edge was generally straight. Another transom, generally the third from the top, was known as the deck transom, because it was arranged so that its upper edge was in line with the planking of the deck, and helped to support it. The other transoms were known as filling transoms.

The shapes of the transoms altered from the higher to the lower. Passing from one to another, the slight curve of the wing transom developed a point where it met the sternpost, and the lowest transoms became closer to an equilateral triangle in shape, though with a slight concave on the outer sides. The transoms also had their outer section shaped so that the lower edge was cut well in from the upper edge, producing a gentle curve over the whole assembly.

The 'fashion pieces' were part of the whole frame in the sense that they were the last of the vertical timbers of the main structure; but they belong in this chapter because they also formed an essential part of the stern assembly. There were usually two fashion pieces on each side on a ship of this size and period. One, the aftermost, began in the deadwood near the end of the keel, and continued upwards to meet one of the higher transoms, where it ended, with the outer ends of the lower transoms being joined to it. The other fashion piece was fitted just forward of this one and carried on to the top of the side of the ship, meeting the ends of the three or four upper transoms. In practice, only the middle parts of the fashion pieces were fitted at this stage in construction. The lowest futtocks could not be fitted, as the deadwood to which they would be bolted was not yet in place. The upper futtocks of the second fashion piece were not fitted until much later in the construction and

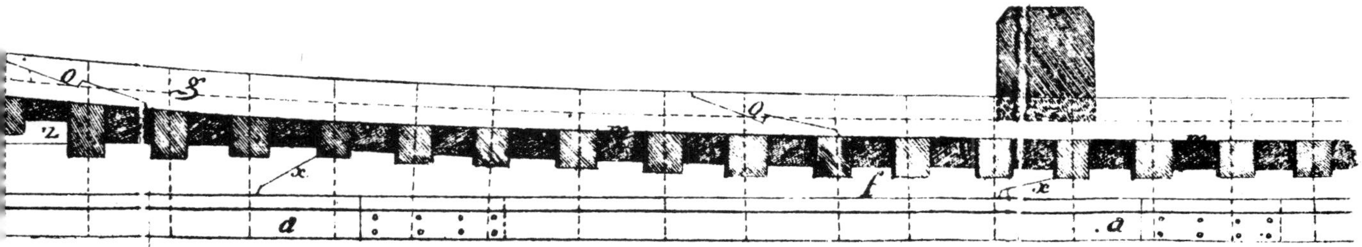

nothing was assembled above the level of the wing transom.

By this time it was becoming less common to erect the sternpost on its own. Instead, the whole assembly of transoms, sternpost and fashion pieces was put together on the dockside and raised as a unit, with the aid of a block and tackle attached to the aftermost sheers. This clearly had certain advantages, as the structure was easier to assemble off site.

The Deadwood

The basic structure of keel, stempost and sternpost was now complete, but it was supported by another structure inside it, and in the same plane, known as the deadwood. As well as adding strength to the whole assembly, the deadwood played an essential role in positioning and securing the timbers of the frame. In midships, the deadwood was relatively simple. It was straight timber, placed on top of the keel and slightly narrower than it. It was cut with large grooves, into which the floor timbers of the frame fitted. In a sense it would have been simpler to bolt the floor timbers directly onto the top of the keel, but in that case there would have been nothing but the bolts to prevent the timbers from working back and forward. The deadwood in midships made for a much stronger join.

Towards the bow the deadwood began to change its form and curve upwards, but its function was still unchanged. From a few feet behind the end of the keel, the deadwood rose up to meet the stempost and the forward frames were not placed across it but the end of each (usually a cant frame fitted at an angle of less than 90 degrees to the keel) was bolted to its side. This part of the deadwood was made in several pieces, the lower part continuing in the line of the deadwood in midships but gently curving upwards. Above it were fitted additional pieces, scarphed together.

In the stern the same principles applied, but the after part of the ship tended to be sharper underwater than in the bow and, as a result, the deadwood at the stern was more extensive than in the bow. The upward curve of the deadwood began soon after the midships point and reached a much higher level. Even forward of the higher deadwood where the cant frames were joined it needed two tiers of timber, each scarphed together. Further aft it was made up of at least three tiers. The longest of these three tiers was the 'knee of the stern' which filled the angle between the false sternpost and the keel. However all the pieces of deadwood formed an even surface when fitted (apart from the recesses to take the frames to be fitted to them).

Once this basic structure – consisting of keel, stempost and sternpost, transoms, fashion pieces and deadwood – had been erected, the first stage of construction was complete. The ship would soon begin to take on a more three-dimensional form, for the timbers which would give the hull its shape were already being cut and brought to the dockside.

8

The Frame

The frame of a ship was its skeleton, formed with ribs, and with the keel in the place of the backbone. But it was more than that, for the structure under the planking was much more solid than the human ribs under the skin. The frame timbers were close together and the structure was about two-thirds solid under the planking. It could withstand a good deal of enemy shot without fatal damage and ships were seldom sunk in action. But despite the massive construction, it had a serious weakness. As Robert Seppings was to point out at the beginning of the nineteenth century, the planks and timbers were all set at right angles to one another, and there was no attempt to use diagonals to brace the structure. Shipbuilders ignored the principle 'well known to the meanest mechanic', that a triangle is much stronger than a rectangle.

Ribs and Futtocks

The skeleton of the ship was a series of ribs, known to shipwrights as timbers, frames or frame bends, and a ship the size of the *Valiant* needed about seventy. It was impossible to find wood which was long enough, or suitably shaped, to make a single rib even on one side of the ship. As a result, each was made up of several sections known as floor timbers and futtocks. The frame bends were placed in pairs, side by side, and arranged so that the joins of the futtocks on one of the frame bends were as far as possible from those of the other.

Looking at one of the bends of a pair, the lowest part was the floor timber, which was about 25ft long in midships, and was the only large piece of wood to pass across the keel, having its centre notched into the deadwood. A floor timber in midships was relatively flat, with its outer ends curved slightly upwards by the floor sweep; but those closer to the bow and stern tended to angle more sharply upwards, demanding carefully selected timber.

The next timber of this bend, the second futtock, joined the end of the floor timber and was fixed to it by means of a chock and bolts. It continued the upward rise of the line of timber, taking in part of the floor sweep and the whole of the reconciling sweep. Then came the fourth futtock, which was shaped to the line of the breadth sweep, the dead flat and the above breadth sweep. Finally, came the toptimber, which mostly followed the line of the toptimber sweep. All the timbers, except the floor timbers, were made in pairs, so that one side of the ship was the mirror image of the other.

The other bend of the pair had a different system of futtocks. It began with the first futtocks, which met at the centreline of the keel and were held together by a long chock over the top of the keel. Each first futtock took in the shape of the floor, floor sweep and part of the reconciling sweep. Next came the third futtock, joined by means of chocks and bolts in the usual way, and following the shape of the hull almost up to the maximum breadth. Finally, there was another toptimber, rather longer than the one in the adjacent frame.

Systems of Framing

There were two main methods of joining the pairs of bends together. In one system the pairs were arranged so that they touched all along the join and were held together by bolts running fore and aft. In another, the two bends were placed slightly apart, and small pieces of wood, also known as chocks, were fitted at intervals between them, with the bolts running through them. A further variation, sometimes encountered around the middle of the eighteenth century, had some of the timbers curved fore and aft as well as across the ship. The purpose of this is rather mysterious, as it would have been wasteful of timber. However, it may have been an attempt to solve a perennial problem; that of arranging the timbers so that they formed the sides of gunports.

The timbers became smaller as they rose up the sides of the ship, both in thickness, or 'sided' dimension, and in depth or 'moulded' dimension. The sided dimension was often reduced in steps, as each timber was made slightly narrower than the one below it. The moulded dimension tapered gradually and smoothly, in order that the hull could be planked both inside and outside. A floor timber in midships of a 74-gun ship was 15½in wide, and it was 13½in moulded at its head. A second futtock was about 1in less in its

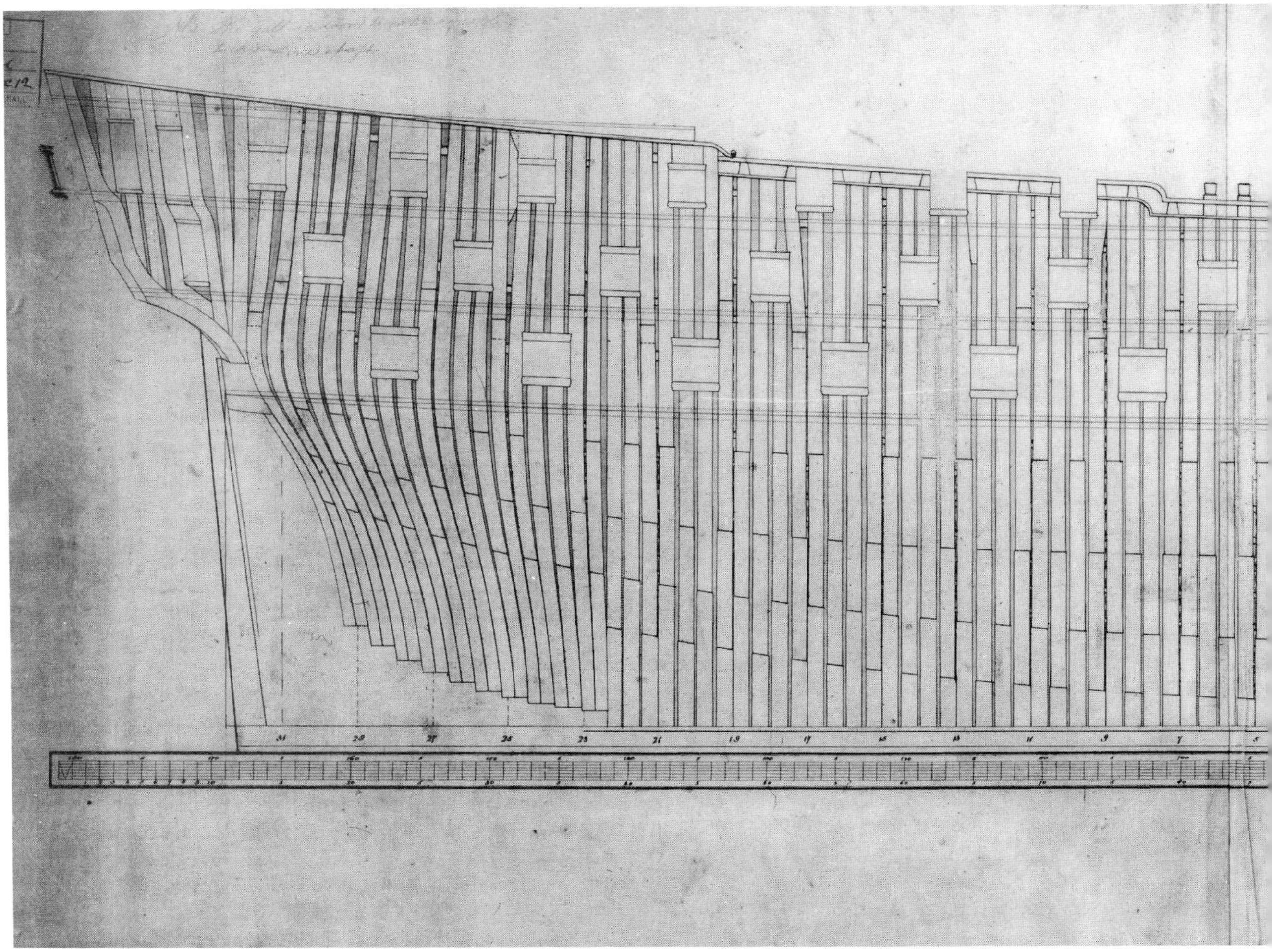

The framing plan of a 74-gun ship. Such plans became common in the 1770s. (National Maritime Museum)

siding, and a toptimber was a further inch less than that. In moulded dimensions, the depth of the floor timber in midships (including the chock) would be about 2ft deep. It was about 13in wide where it joined the second futtock, and at the level of the upper deck the moulding of the timbers had further declined to 10½in. At the top of the side in midships, the toptimber was about 5in moulded, and towards the stern, where it carried up further to cover the quarterdeck and poop, it was further tapered to 4½in.

The 'room and space' was the distance, measured fore and aft, between a point on one pair of bends and the equivalent point on the next. For a 74-gun ship of 1779, this distance was 2ft 8½in, and for the *Valiant* it was 2ft 8in. Taking the ship of 1779 as an example, the siding of the floor timber and first futtock at the keel added up to 2ft 6½in, leaving only 2in between the timbers. At the level of the upper deck the siding of the two timbers amounted to 21in, so the hull was still about two-thirds solid. At the top of the side in midships it was rather less so, for the combined siding of the two toptimbers was only 10in, leaving a space of 22½in.

The Cant Frames of the Bow

Towards the bow and stern, the system of framing the hull changed radically. In the central part of the ship the timbers were placed across the keel at right angles to its line. Towards the extremities of the ship the hull narrowed and so the planking no longer ran parallel to the line of the keel; indeed, up in the bow it was almost at right angles to it. Furthermore, the frames rose more sharply from the keel as they approached the bow and stern, and it became increasingly difficult to find timbers with a suitable grain. The solution,

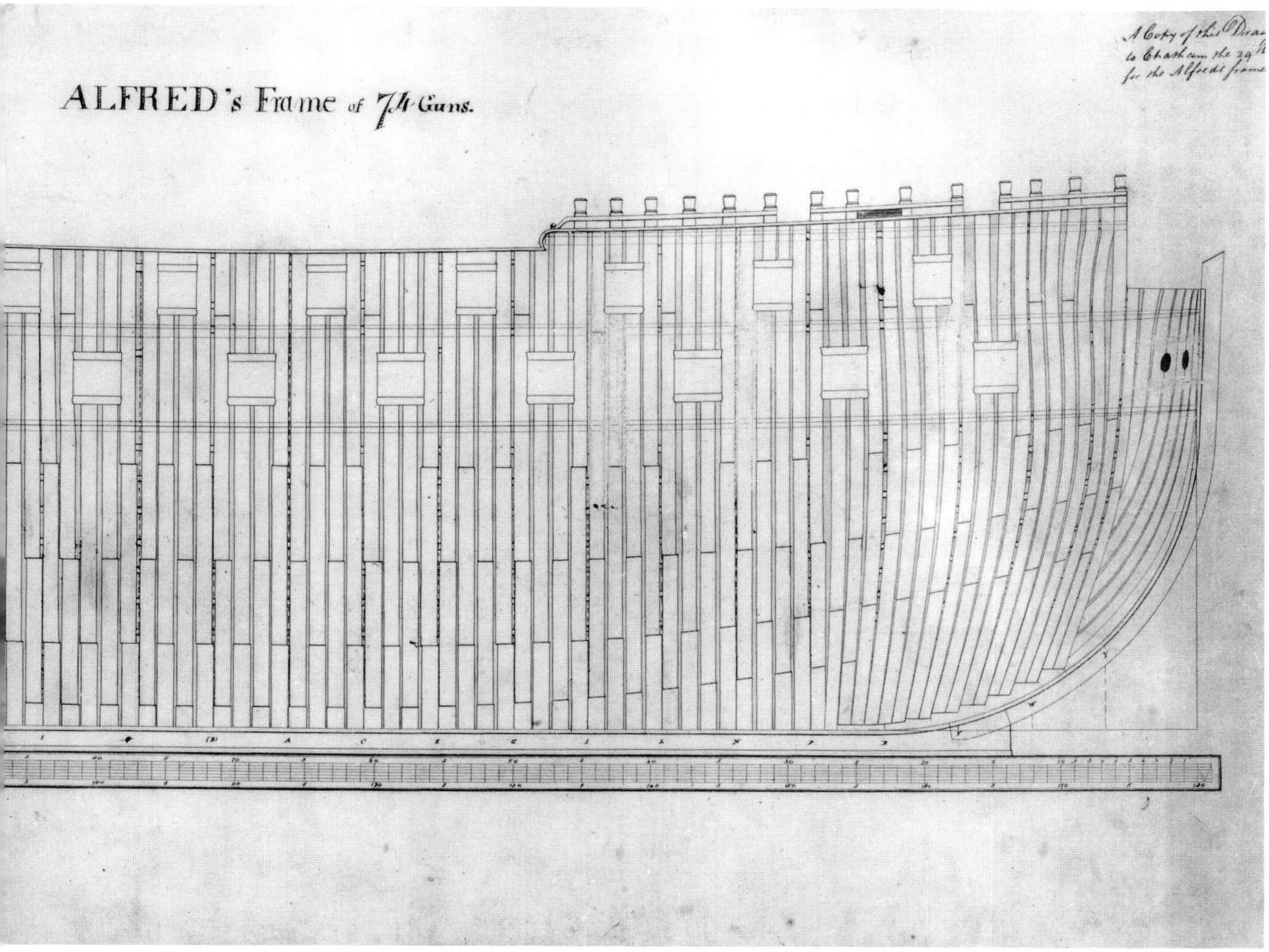

adopted officially for the navy following a report of 1715, was to use 'cant frames' instead of 'square frames' near the extremities. Forward of frame S on the *Valiant*, about 25ft from the stempost, cant frames were fitted. The aftermost one was only at a slight angle to the square frames. This angle increased with each frame, so that the foremost one was fitted at an angle of about 30 degrees. Clearly a floor timber could not be used as the lowest part of a cant frame; it would have to bend in both the horizontal and vertical planes, and it would not have been easy to find timber with suitable grain. Furthermore, the timbers close to the bow and stern rose sharply upwards from the keel. Therefore, the cant frames, and perhaps a few of the square frames close to them, had 'half timbers' rather than floor timbers. The half timbers, as their name implies, did not pass over the keel, but were made in two halves which joined above the keel; they were similar to first futtocks, but rather shorter. In the area where half timbers were employed, the deadwood rose some way above the keel, and the ends of both the half timbers and the first futtocks were butted into it and bolted across the keel. Above that rose a normal structure of second, third and fourth futtocks, and toptimbers.

Hawse Pieces

The bow of a large ship of this period was very bluff, and about 6ft aft of the stempost the line of the hull at the waterline was about 60 degrees to the direction of the ship's forward movement. The system of cant frames ended there, to be replaced by timbers known as 'hawse pieces', because they contained the hawse holes through which the anchor cable passed. The hawse pieces, five to six on each side in a 74-gun ship, all ran forward from the first cant frame and in the side plan view they were parallel to the line of the keel. The bollard timber was right forward and being the innermost was bolted directly against the stempost and

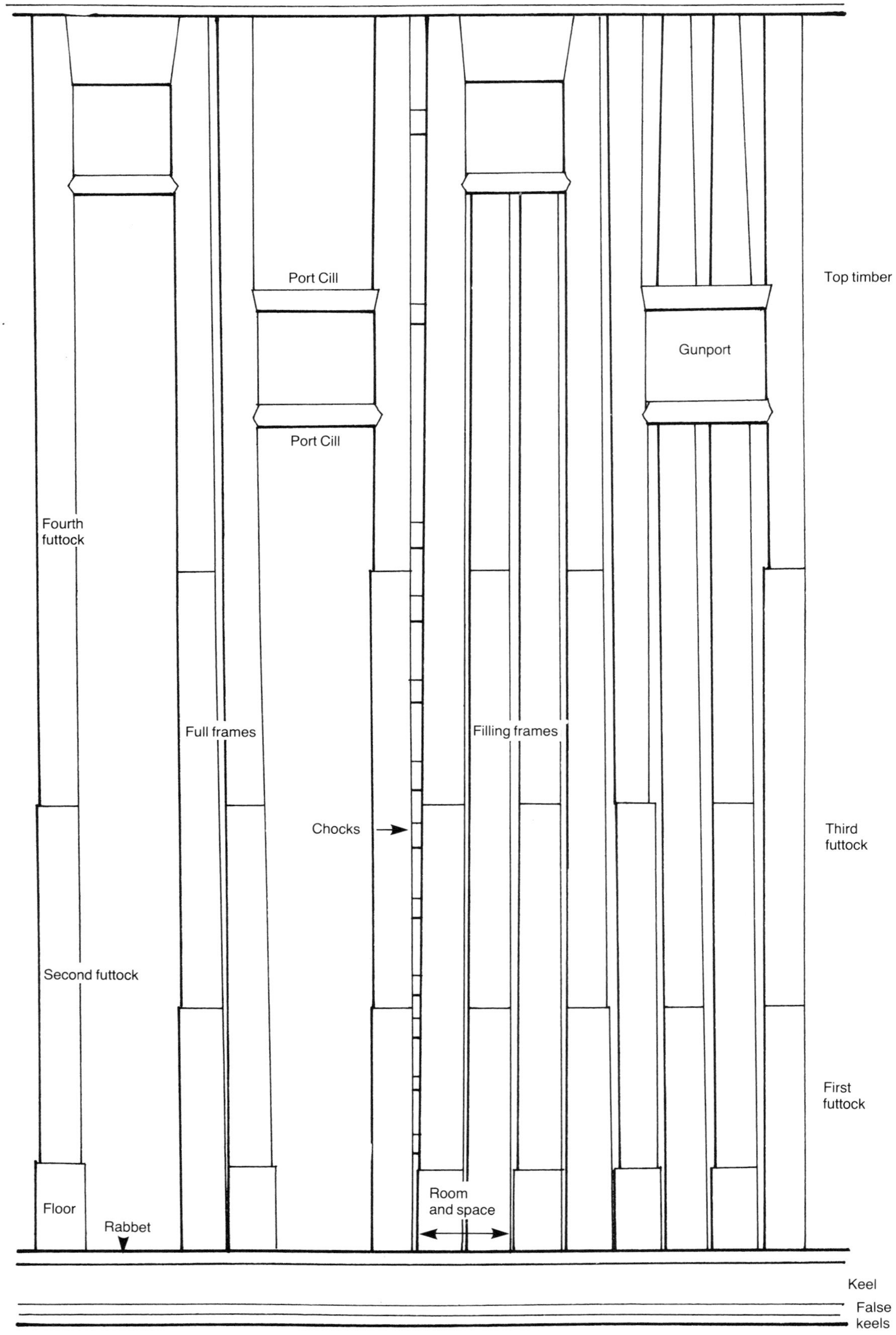

Port Cill
Top timber
Gunport
Port Cill
Fourth futtock
Full frames
Filling frames
Chocks
Third futtock
Second futtock
First futtock
Room and space
Floor
Rabbet
Keel
False keels

The basic layout of the frames in midships. The first drawing shows the method with the pairs of frames separated and the second shows them touching.

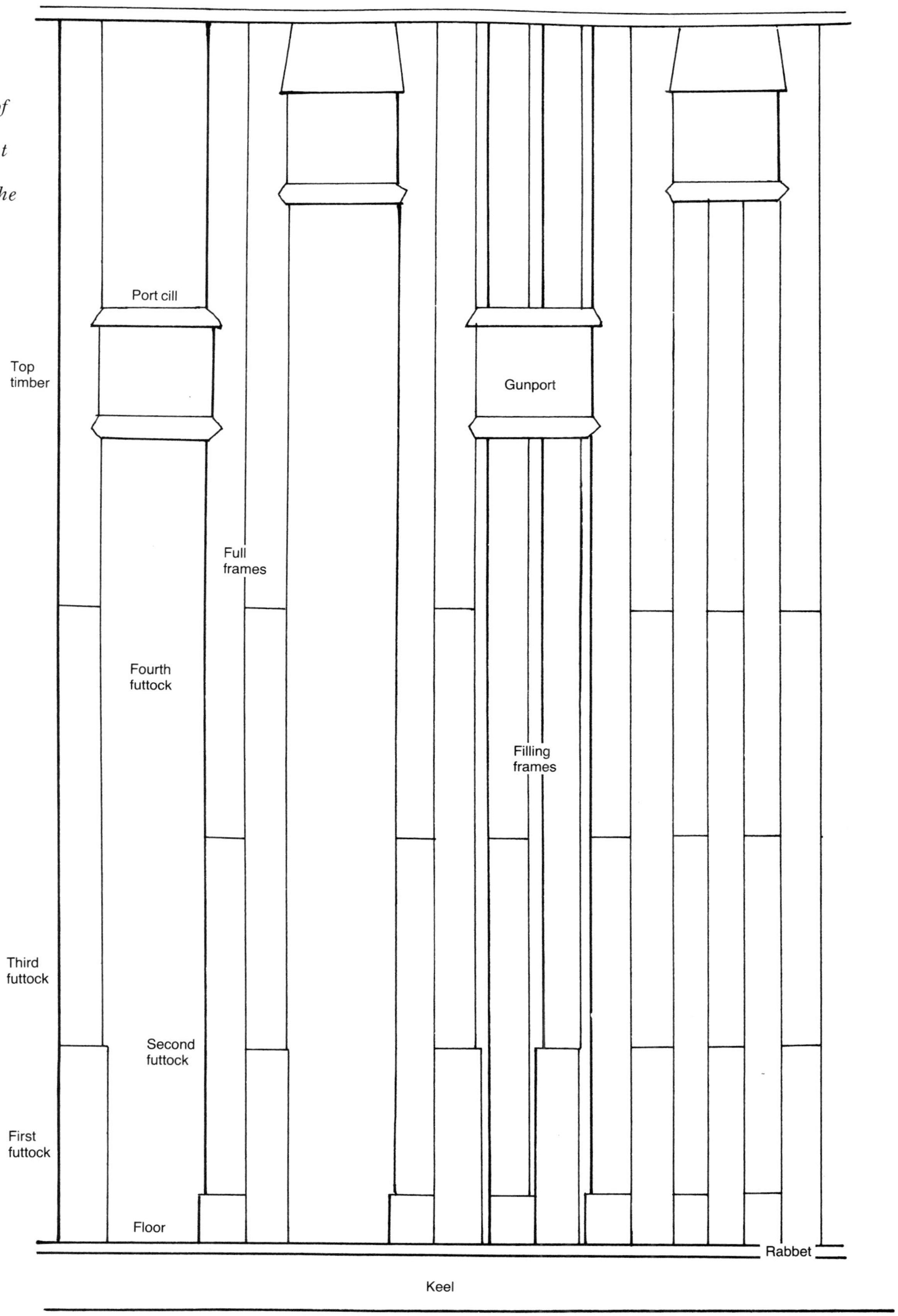

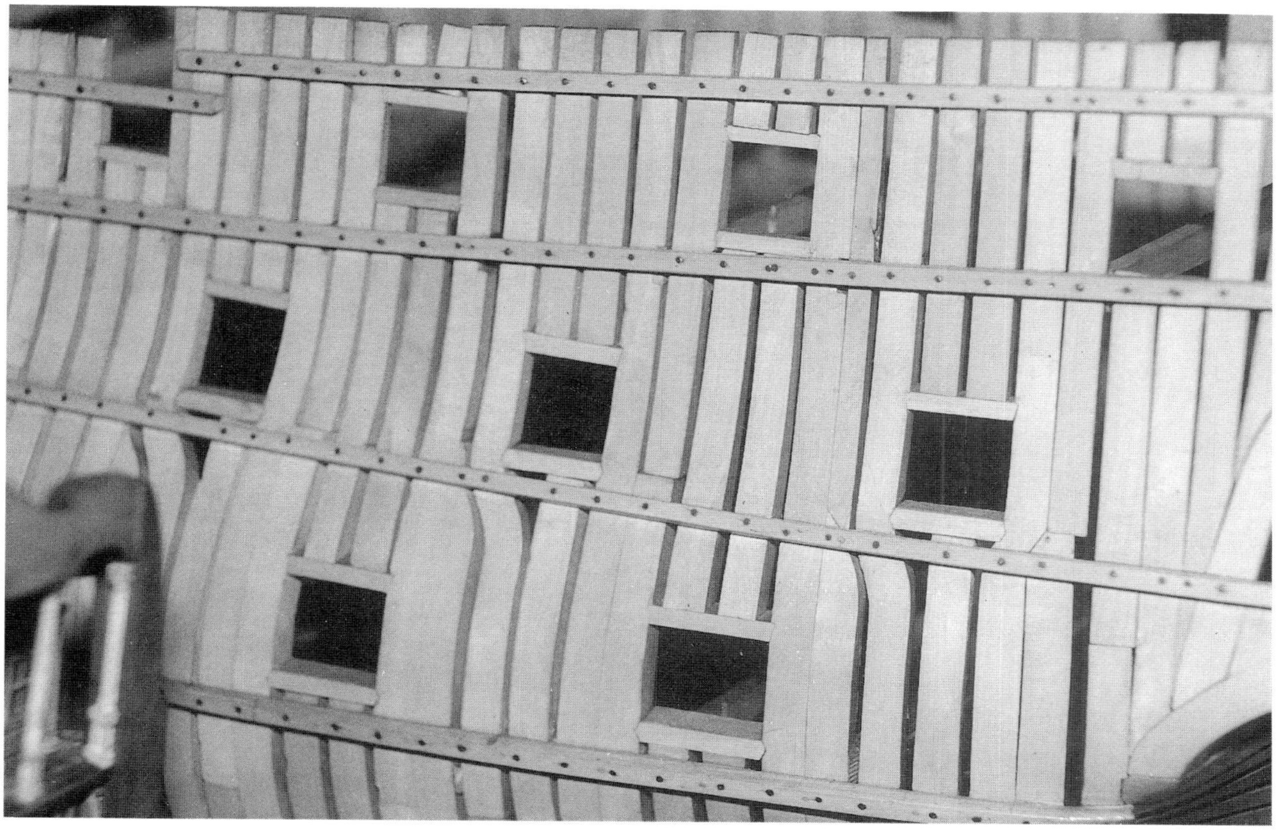

The unusual framing of the Bellona *model, with some of the frames curved backwards or forwards to form the sides of the gunports. (Author's photograph, courtesy of the National Maritime Museum)*

◀ *The outside and inside of a model of the early nineteenth century showing the system of framing. The central part shows the full frames fitted, with the filling frames in place on either side. The chocks have been left out, and this shows clearly the joins of the different futtocks. (The Science Museum)*

knee of the head. It carried on much higher than the other hawse pieces, and even above the level of the knee of the head. It ended in the 'knight heads'. One of these was fitted on each side and they served to brace the bowsprit against lateral pressures. In some cases, the bollard timbers may have been fitted earlier in the construction, before the rest of the frame timbers. This could have been useful in holding the stempost and knee of the head in place.

The Stern

The stern too had a system of cant frames, beginning about 25ft from the sternpost and these ended in the horizontal transoms rather than the vertical hawse pieces. The fashion pieces, fitted at an earlier stage in construction, were in fact the last cant frames, and were faired with them when the structure was set up. Under the lowest transom were more short vertical timbers, known as filling pieces which ran down into the deadwood.

Above the wing transom, the structure of the stern was light. Its after face was formed by six timbers, known as the counter timbers, rising from the wing transom. Seen from the side, they were similar in form. Each started off with two curves, which formed the shapes of the two counters. Above that the timber was straight, and angled backwards to form the flat surface of the stern. From the stern, the four inner counter timbers were straight, and angled inwards as the stern narrowed as it rose from the transom. These timbers ended at the level of the quarterdeck. The remaining pieces were known as the side counter timbers and rose from each end of the transom forming the corners of the stern. Seen from the stern, each had an outward curve on

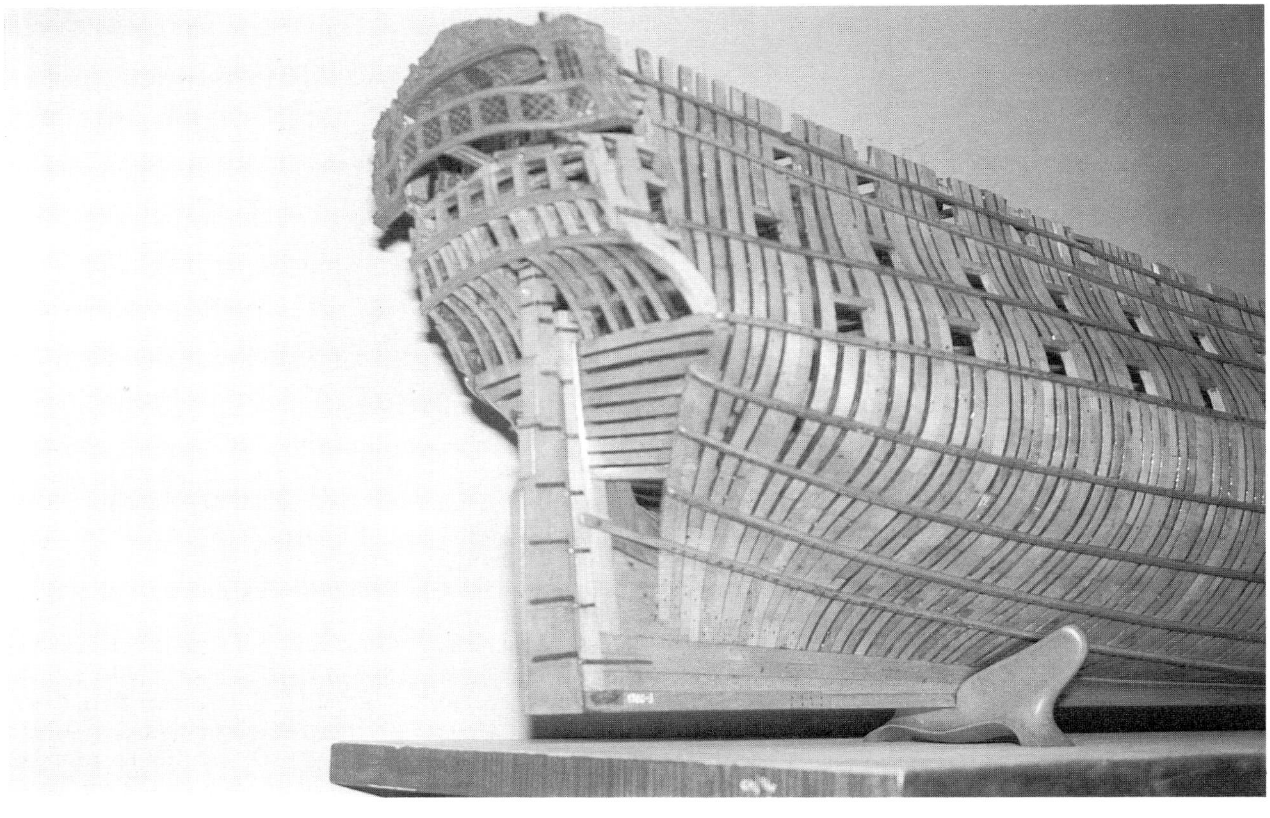

The bow of the Bellona *model, showing the square frames, cant frames and hawse pieces. (National Maritime Musuem)*

A ship under construction at Chatham, with the keel laid, the stempost raised and the floor timbers crossed. (National Maritime Museum)

its lower part, comforming to the shape of the hull where the main wale met the stern. After that each was straight, and continued all the way to top of the side.

Short counter timbers were fitted between the main ones as required, partly to form the sides of any gunports that might be fitted in the counter. Above the level of the quarterdeck, the structure was even lighter, and formed the aftermost side of the captain's cabin. It consisted of removable wooden partitions, with numerous windows. Even below this level the structure was relatively light, nowhere near as strong as that of the sides of the ship, and the stern was the most vulnerable part of any warship of this period.

Between the side counter timbers, which formed the corners of the stern, and the fashion pieces, another system of framing was used to form the sides of the ship. At the upper levels, it was similar to the rest of the structure of the sides, with toptimbers and other frame timbers. They were interrupted by doors to the quarter galleries, instead of gunports. Their lower ends joined to the side counter timbers, placing considerable strain on these parts, which in turn was transmitted to the wing transom. Naturally, the timbers of this part of the stern got much shorter as they progressed aft, because of the angle of the counter timbers.

Raising the Frames

The actual assembly of the frames was done in a particular order. The floor timbers and half timbers were fitted first, laid across the keel or butted into the deadwood of the bow and stern. When this was finished, a merchant shipbuilder was entitled to the second instalment of his payment (the first being paid on the signing of the contract). Meanwhile, the upper parts of the frames were being cut by the shipwrights, and put together by the dockside. The whole frame bend for each side – first, second, third and fourth futtocks and toptimbers – could be bolted together. When completed it was raised into place as a unit. A block and tackle was slung at an appropriate point from the rope stretched between the sheers at each end of the hull. According to pictorial evidence, both sides of a frame were raised at the same time, perhaps so that the two tackles would balance one another and keep the rope reasonably central.

The first frames to be raised were the 'full frames'. These were the ones which were to be uninterrupted by the gunports and they were carefully arranged so that they would eventually form the sides of the ports. There was a

◄ *The stern of the* Bellona *model. (The National Maritime Museum)*

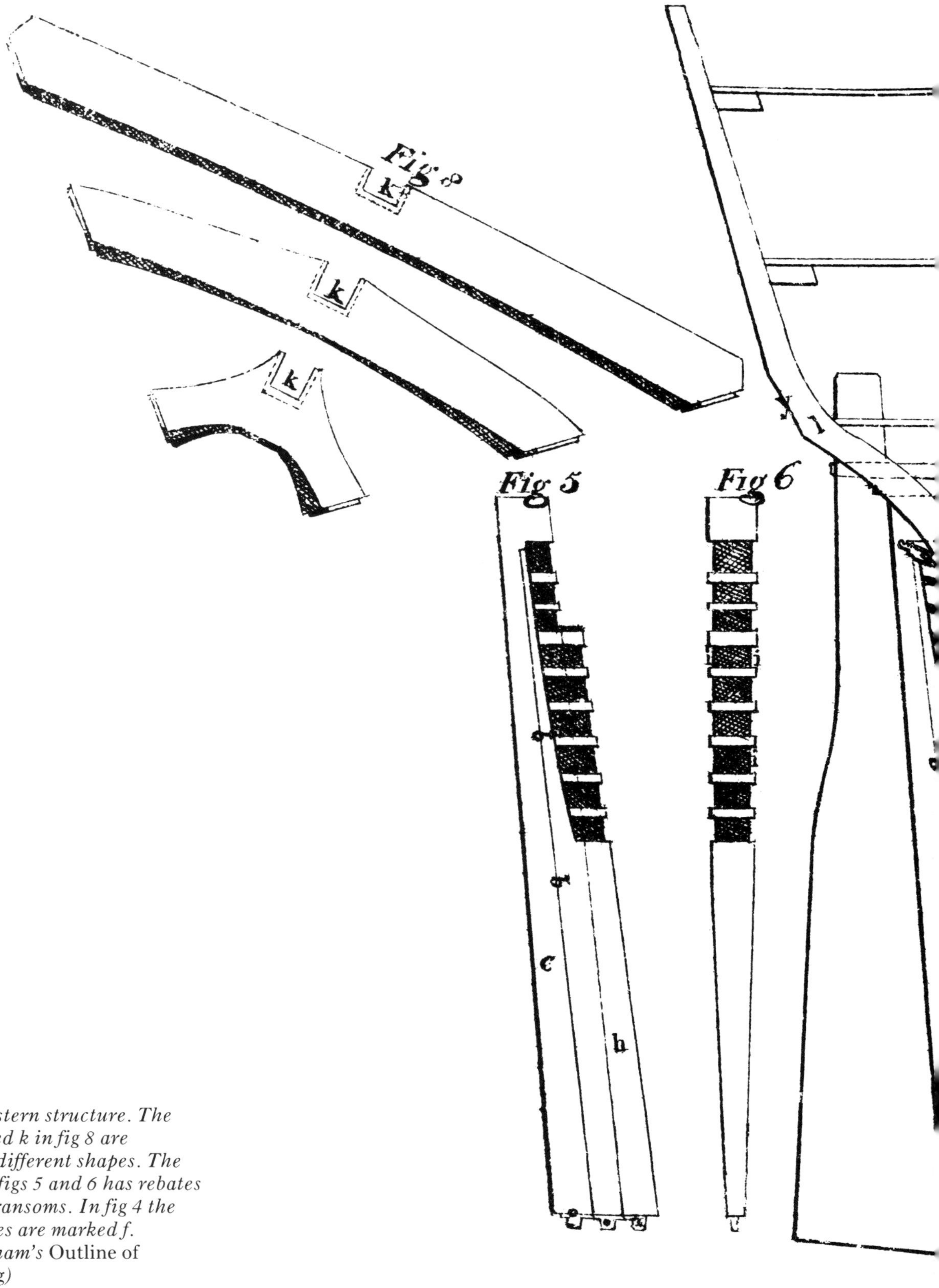

Parts of the stern structure. The pieces marked k in fig 8 are transoms of different shapes. The sternpost in figs 5 and 6 has rebates to take the transoms. In fig 4 the fashion pieces are marked f. (From Fincham's Outline of Shipbuilding*)*

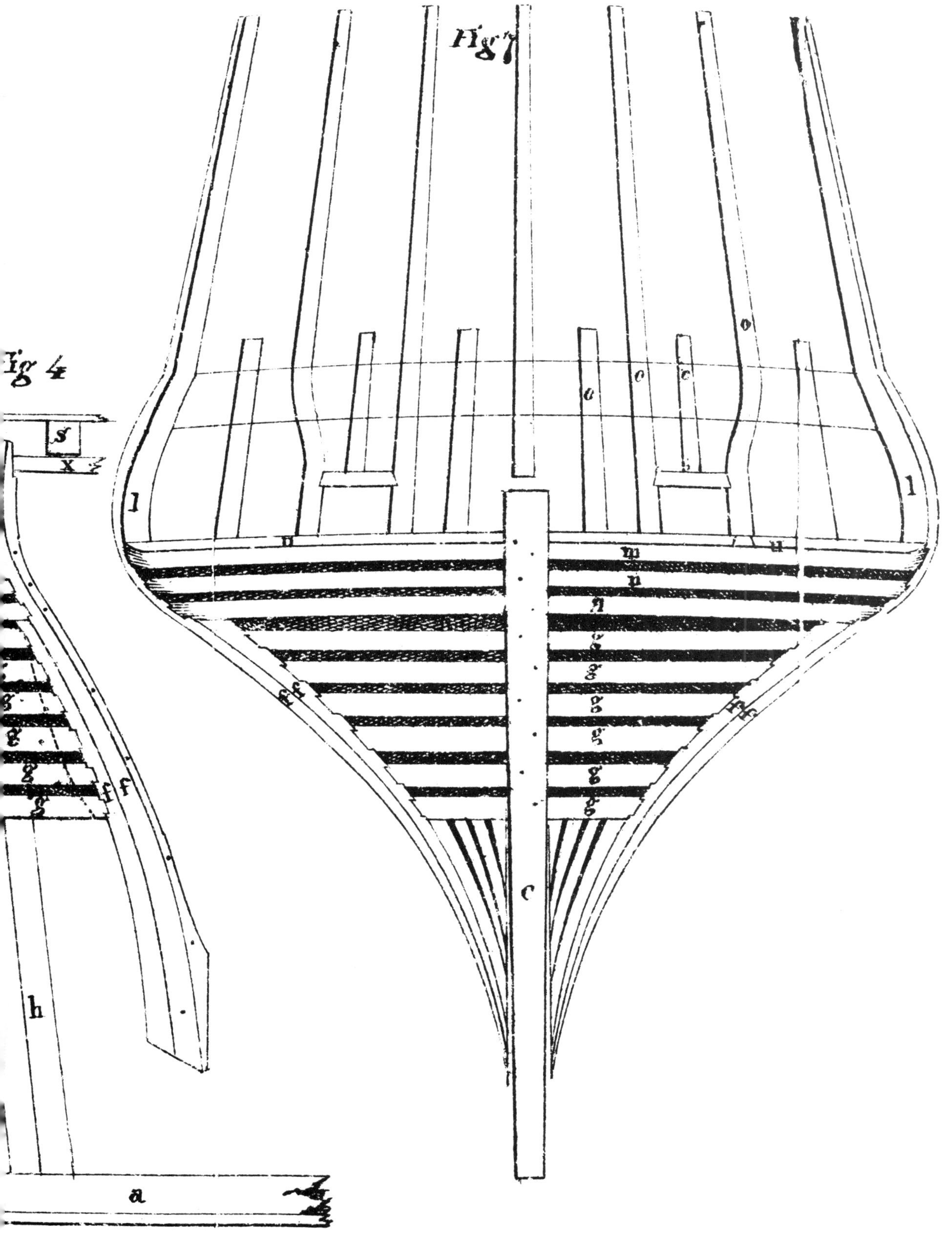
Fig 7
Fig 4
s
x
l
l
u
u
m
p
o
o
o
o
c
h
a
f f

Raising the frames using the sheers and an arrangements of blocks and tackles.
(British Library)

tendency for ships to be built from the midships frame forward and aft, but much depended on when the individual timbers became available. After that, the port cills were fitted. They formed the top and bottom edges of the ports, and were fitted between the full frames at the appropriate heights, with their ends recessed slightly into the frames. The port cills matched the thickness of the timber at that point, and on the lower deck of a 74 they were 7½in deep. The 'filling frames' were fitted next, between the full frames and above and below the gunports. Often the chocks between the futtocks were not fitted until later in the construction, to allow the timbers to season better.

When the assembly of frames was reasonably complete, the keelson, or kelson, was placed over them. This timber was similar in thickness and depth to the keel and like it was made of several pieces scarphed together. It fitted over the floor timbers and first futtocks, directly above the keel itself, and helped to lock the assembly together. Towards the bow and stern the keelson served simply to strengthen the deadwood to which the cant frames were attached.

The Knee of the Head

The knee of the head was fitted after much of the framing was complete and was placed directly forward of the stempost in the same plane as the keel, stempost and sternpost. It served to support such decorative features as the figurehead and rails, and also provided a place to fix the gammoning ropes which held the bowsprit in place, and for the bobstay which secured the outer end of the bowsprit. The knee of the head was shaped like a bracket, with its outer end turned sharply upwards to provide a seat for the figurehead.

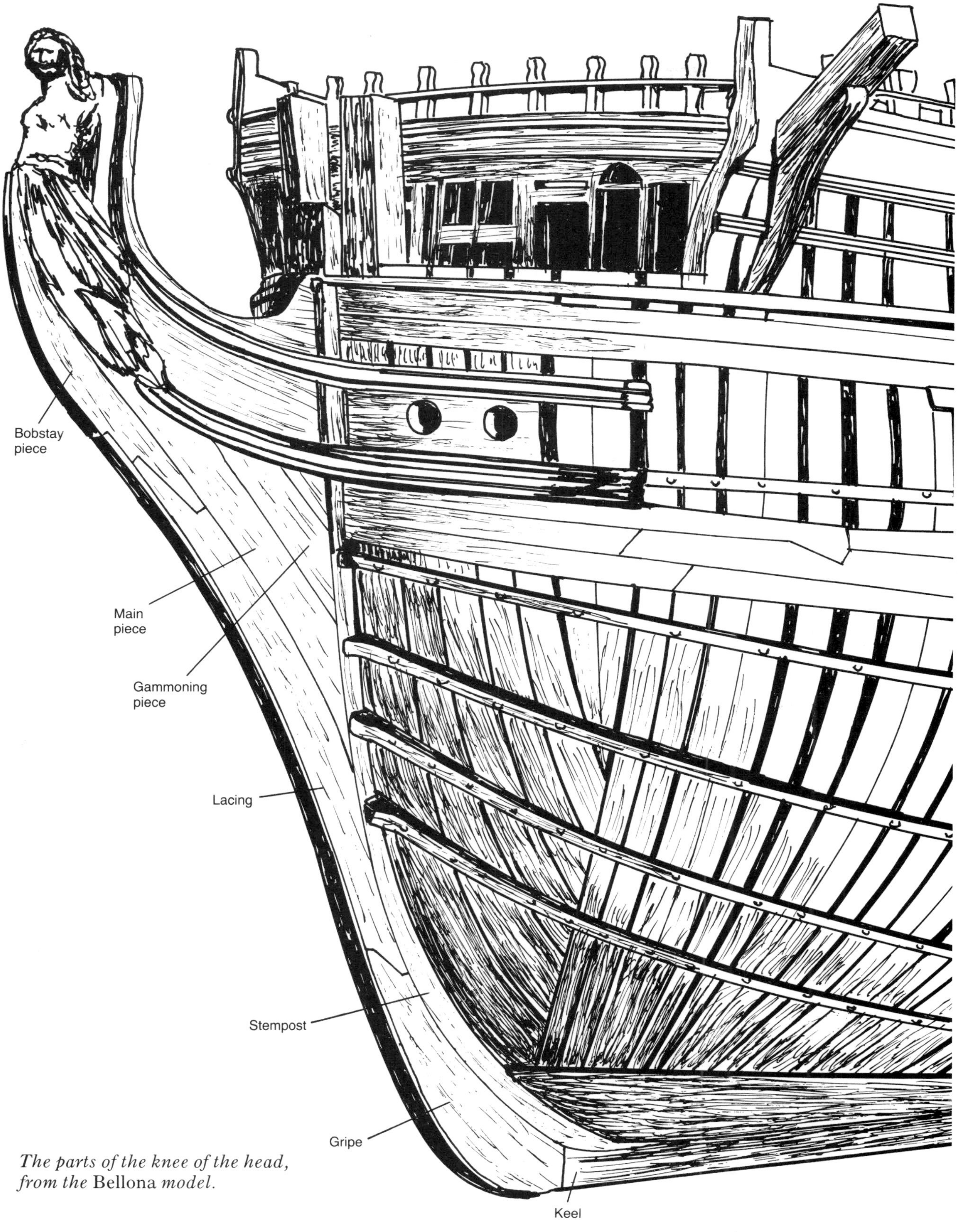

The parts of the knee of the head, from the Bellona *model.*

The Victory *under construction with scaffolding raised around it. In fact, the* Victory *was built in a dock rather than on a slip. (Author's photograph, courtesy of the National Maritime Museum)*

Its lower end joined the forward end of the keel and was known as the gripe; above it was the lacing, which carried the outer shape up almost as far as the lowest part of the figurehead. From the middle of the lacing upwards, the knee of the head was too wide to be made in a single piece. Behind and above the lacing was the main piece, which provided much of the strength of the structure, and supported the figurehead directly. Behind that was the gammoning piece, which had holes for the gammoning ropes of the bowsprit. The bobstay piece was fitted forward of the main piece, as a continuation of the line of the lacing. It was the part whose outer edge curved upwards to give the head its distinctive shape, and it contained the holes for attaching the bobstay. The whole structure was held together by a system of scarphs and coaks and the horseshoe plate, so called because of its shape, was bolted on the outside to attach the gripe to the lower part of the stempost.

Supporting the Frames

During this stage of construction the frames were not bound together by the planks or the deck beams as they would be in the finished ship; furthermore, the whole assembly lacked the support of water which it would have after the launch. Three systems of temporary support were therefore employed during the period of construction.

In the first place, the frame timbers of each side of the ship had to be held together and kept vertical. This was done by means of ribbands and harpins. Ribbands (or rib-bands) were simply 'long narrow flexible pieces of timber, nailed upon the outside of the ribs'.[1] Three rows of ribbands were fitted to the hull of the *Valiant* below the maximum breadth and this was the case with most ships of her size. Their positions were carefully planned on the draught of the ship, and they appeared as diagonals on the body sections drawings – one at the position of the floor timbers and therefore known as the floor ribband, another through the reconciling sweep and a third passing through the below breadth sweeps. The position where each ribband met each individual frame was drawn on the moulds, and then marked on the futtocks as they were cut out. These marks, known as surmarks, ensured that the structure was fitted together properly. The ribbands also served another purpose; they helped to ensure that the lines of the hull were truly fair. Any miscalculations on the part of the designer, or bad workmanship by the shipwrights, would produce local irregularities on the frames and the ribbands would accentuate these. The hull could then be 'dubbed', the shipwrights working with their adzes 'to square the outer sides so that the plank should lay fair and even upon them'.[2] Sometimes this was done after the hull had been allowed to season.

At the bow and stern, the shape of the hull was even more crucial. In these areas the flexible ribbands were replaced with carefully shaped pieces known as harpins. When fitted they looked rather similar to ribbands, but in practice they were cut from solid timber, rather than bent in situ, and the timbers were dubbed to fit them.

The second system of temporary bracing was internal. The deck beams would eventually connect the two sides of each frame, but they were not fitted at this stage and so 'cross spalls' were used instead. These were comparatively light pieces, fitted both horizontally and diagonally in order to maintain the shapes of the frames. Finally, shores were used outside the hull to keep it upright in the dock. The first shores were fitted to the stemposts and sternposts and rose almost vertically from the dock to hold the pieces in place. More shores were placed along each row of ribbands, rising diagonally from the steps on the side of the dock, and fitting under the respective ribbands and harpins.

Throughout most of the age of sail it was agreed by the best authorities that a ship should be allowed to 'stand in frame' for a period, to allow its timbers to season after they had been assembled into the frame of the ship. However, there was no clear policy on this until the 1770s when it was found that the ships which had been hastily built in the Seven Years War were suffering from swift decay and it was made mandatory then that each new ship stand for a year before being planked. It is not clear how long the *Valiant* was allowed to stand in frame, but it was certainly much less than a year; there is some evidence that she stood in frame during the winter of 1758-9. After that, she was ready to have her decks fitted, and to be planked.

A ship under construction at Sheerness with props and cross-spalls in place. (Author's photograph, courtesy of the National Maritime Museum)

9

The Planking

Planking covered the frames of the ship, both inside and outside. It was nearly all made for oak, except that elm was used in selected places below the waterline. In contrast to the frame, shipbuilders were quite happy to use foreign timber, especially 'east country plank' from the Baltic.

Most of the planking was cut in long straight pieces but wales, which needed extra strength, were often made up of short interlocking pieces. The thickness of planks varied, both inside and outside the ship. Externally, the wales and rails were the thickest parts and projected a few inches beyond the normal planking; the planks close to the wales were also slightly thicker than normal though these were faired in with the adjacent planking instead of projecting sharply. All the external planking below the waterline was of an even thickness, to create as smooth a surface as possible. Internally, the planking was thicker over the joins of the futtocks and here there was no attempt to fair it into the surrounding planking.

The Role of Planking

The external planking served two main purposes: to keep the water out of the hull, and to give the structure its longitudinal strength. The internal planking served to keep ballast and debris from falling between the timbers, and itself added much strength to the ship. The planking of the decks, of course, provided a reasonably level and smooth surface for living and working, and it added considerable strength to the structure.

Because planks were mostly cut in straight lengths, some treatment was necessary to make them fit the complex shape of the hull. Clearly they had to be bent inwards, especially near the bow where the curve of the hull was greatest. In many cases they also had to be bent upwards, as the plank had to follow the sheer of the wales. This was done by means of a steam chest, in which planks were treated to make them pliable. Even this was not enough in some situations. Where the wales curved round the bow, it was usually necessary to carve the pieces to shape, as the timber was too thick for steaming. However, wales were made in relatively short pieces and this would have made the process easier.

Strakes and Butts

Planks were arranged in strakes. A strake was defined as 'the continuation of planks joined to the end of each other, and reaching from the stem, which limits the vessel forward, to the sternpost and fashion pieces, which limit her length aft'.[1] A strake, then, was not a single plank, for it would not be possible to find one long enough for the whole ship, but a row of planks, extending the whole length of the ship. The planks met end to end, and such ends were known as butts. Normally there was no attempt to join the butts together, by scarphing or any other means. However, the builder was concerned, as always, to prevent any line of weakness developing in the hull and he made sure that the joins of adjacent planks were spaced apart; to do this he used a system which became known as 'four step butt planking'. Twenty-four feet was regarded as the optimum length for a plank, and 6ft was needed between the butts of one strake and those of the one above. As a result, the positions of the butts repeated themselves on every fourth strake; those of the next strake were 6ft along, so that the butt ends rose like steps.

The ends of the strakes near the bow and stern created a different sort of problem, especially in the bow. Those near the waterline, for example, met the stempost almost at right angles, and were liable to spring out because of the great curve in that area. The rabbets of the posts were carefully shaped to keep them in place, but these areas, known as the 'hooding ends' of the planks, were always vulnerable. After she ran aground in 1758, the *Invincible* was regarded as being in serious danger after her hooding ends began to spring out.

Wales

The wales were often the first parts of the planking to be fitted, and in many respects they differed from the general planking. They were of course much thicker, the main wales of a 74-gun ship were 8½in thick compared with 4in for the surrounding planking. In addition, the system of construction was very different. Until about 1715, British ships had had double wales, each

The old type of wale with two separate planks and thinner planking between them. (Author's photograph, courtesy of the Science Museum)

wale being made up of two separate strakes of thick timber, several feet apart. This system was still used for French ships, including the *Invincible*, but in 1715 it had been ordered that for British warships the wales themselves be made slightly thinner, and the plank between them thicker. The immediate result was that the wale became a single structure, made up of three or four strakes of thick plank placed together. By the 1750s, it was no longer common to use straight planks for the wales. The most common method in Britain was now 'anchor stock construction' whereby the parts of each wale were cut in five-sided pieces, rather like a blunt triangle with the outer corners cut off. The pieces of the wale were carefully shaped so that they would interlock, with each alternate section being turned upside down. Most wales of the 1750s used four strakes which were fitted together using this method. A slightly more complex method, known as 'hook and butt', was also used. Pieces were cut with projections and recesses, so that they would interlock and produce a yet stronger join.

As the wales were fitted the ribbands were taken off, but as they had a much greater sheer than the wales, this usually meant that each ribband was taken off bit by bit as it was replaced by the wales. The wales were arranged so as to avoid, where possible, crossing the gun ports but towards the bow and stern the wales rose quite sharply and it was inevitable that they should cut across the ports. A two-decker like the *Valiant* had two wales on each side – the main wale under the gundeck ports and the channel wale under the upper deck ports. A three-decker had a middle wale added. The wales had considerably more sheer than the decks because they were originally intended to help brace the whole ship in the vertical plane; the beams and standards of the lower deck would be on a level with the wales in midships, while towards the bow and stern the wales would be linked with the hanging knees of the upper deck. It is doubtful whether this did strengthen the ship and there was a tendency over the years to flatten the wales. On a 74-gun ship, the main wale was 8½in thick while the channel wale was 5½in.

Planking of the Bottom

The planking of the bottom began immediately under the main wale and generally followed its shape, to begin with at least. The first strake under the wale was known as the 'black strake', possibly because it had once been the custom to tar it.

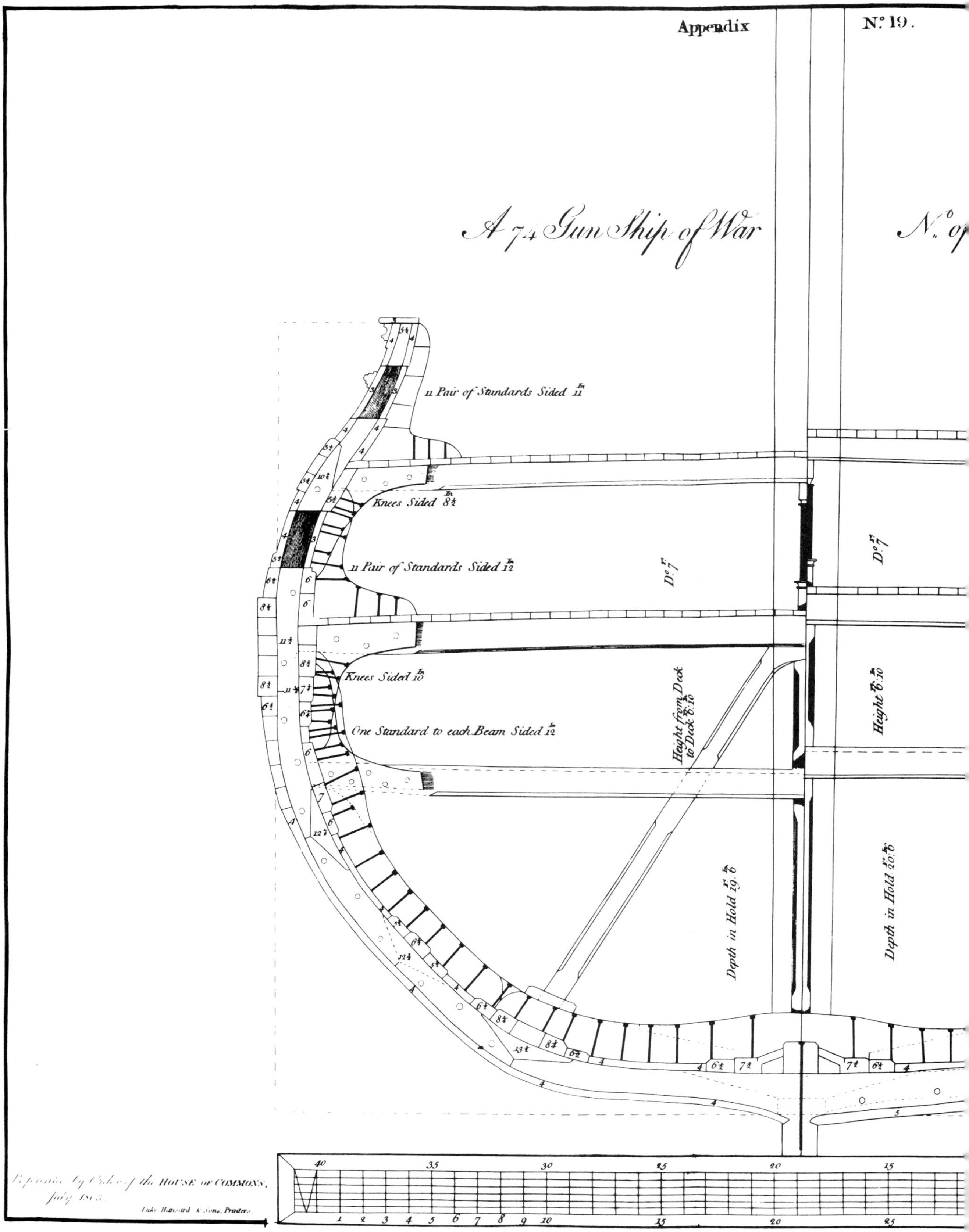
Appendix
N.° 19.
A 74 Gun Ship of War
11 Pair of Standards Sided 11 In
Knees Sided 8½ In
11 Pair of Standards Sided 12 In
D.° 7 Ft
Knees Sided 10 In
One Standard to each Beam Sided 12 In
Height from Deck to Deck 6.10
Height 6.10
Depth in Hold 19.6
Depth in Hold 20.6
HOUSE OF COMMONS,

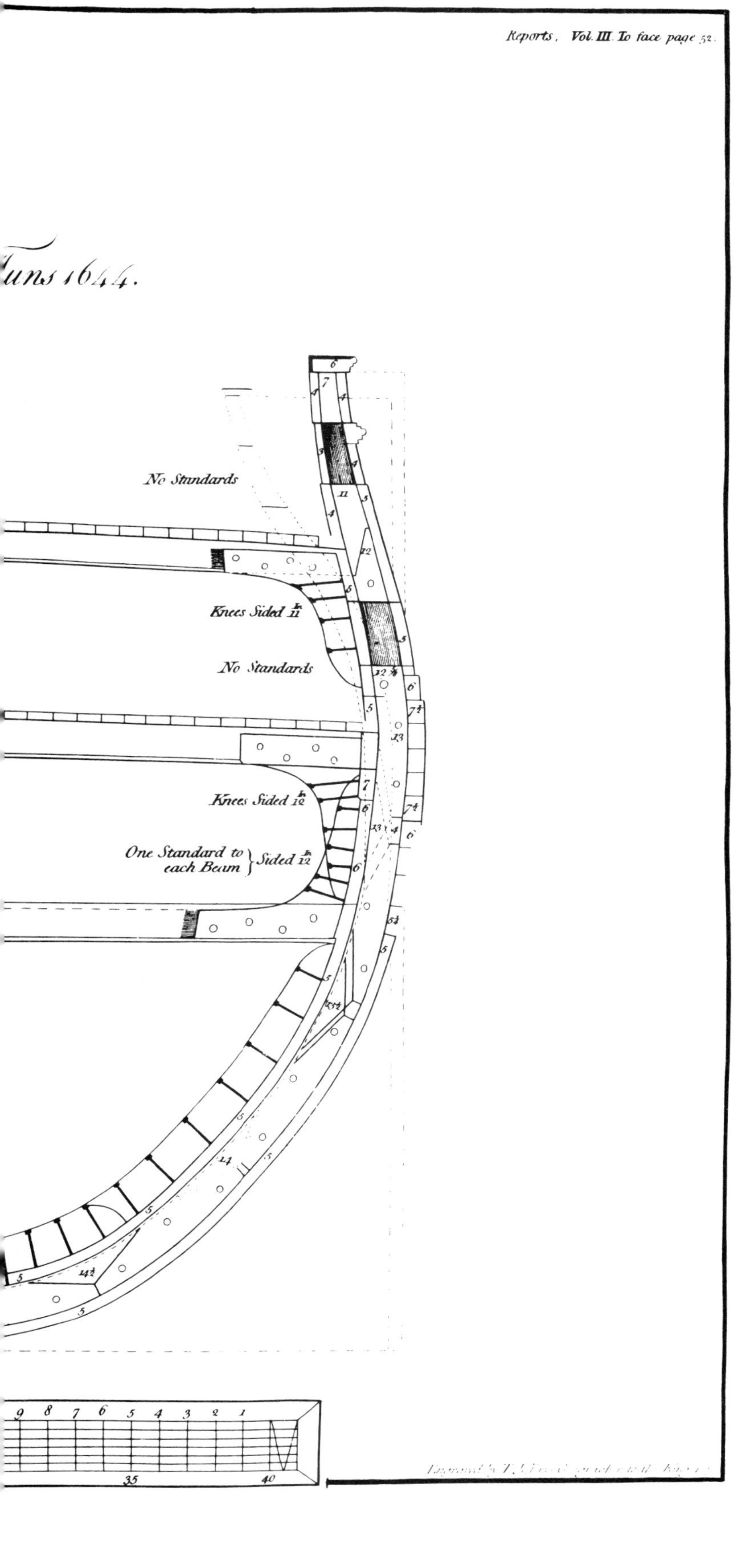

The midship section of a 74-gun ship from a Parliamentary Report of 1771. The right hand side shows projected improvements, but the left side shows the standard method, except for the diagonal brace. It gives clear information on the thickness of plank used, the types of chock, etc. (The British Library)

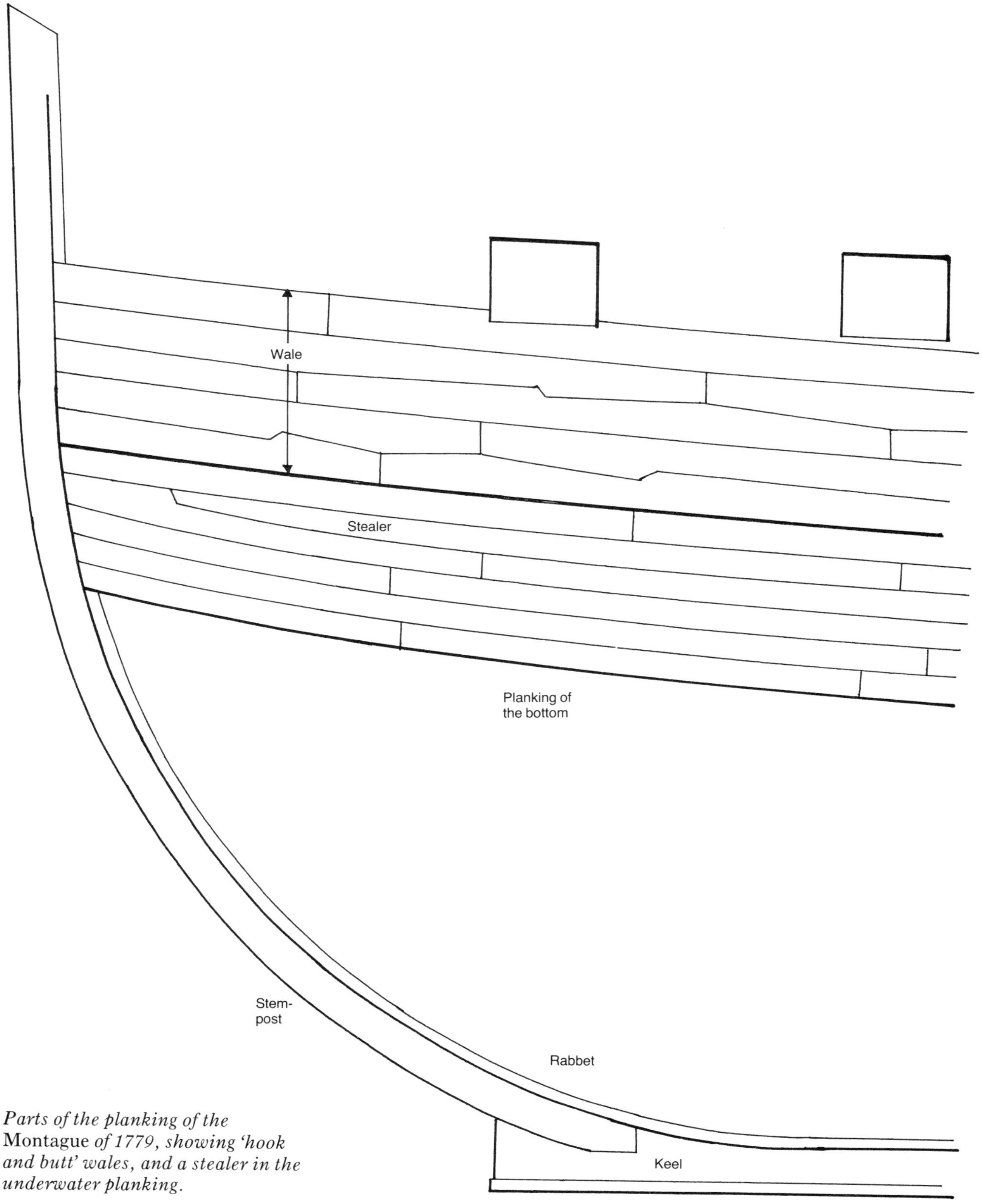

Parts of the planking of the Montague *of 1779, showing 'hook and butt' wales, and a stealer in the underwater planking.*

The wales being fitted on the Bellona *model. The ribband is still in place, and will have to be removed when the next section of wale is put in place.*
(Author's photograph, courtesy of the National Maritime Museum)

It was 6¾in thick, not much less than the wale itself. The next five strakes under the black strake were known as the 'diminishing strakes' and each was progressively reduced in thickness, the whole surface being trimmed with an adze to present a reasonably smooth finish and to fair in with the normal planking below, which was only 4in thick.

Towards the bow and stern the widths of the planking had to be reduced because the hull was reduced in cross section. At the stern, half the planks led up to the counter and the other half to the sternpost. As a result, the widths of the planking did not have to be reduced greatly, and a little tapering was all that was necessary. At the bow the planking had to end at the stempost so the reduction in width was much greater. Planks were tapered as much as possible but usually this was not enough and some planks came to an end before they reached the stempost. Such strakes were a few yards shorter than the others, and were known as 'stealers'. Each was cut to a point at its end and the adjacent planks above and below were cut to shape to fit this.

The very bottom plank, the one which fitted into the rabbet of the keel, was known as the 'garboard strake'. By the early nineteenth century it was normal to make the garboard strake from elm instead of oak, but this does not seem to have been standard in 1759. The contract of 1779 specified that the lowest seven strakes on each side of the keel should be left out during construction to allow chips to fall out and rain water to drain away. This may have been part of the measures introduced in the 1770s to counter decay in ships and was probably not done in earlier periods.

The planking was fixed on by trenails, and the contract of 1779 specified that they be wedged at each end. As a general rule, two trenails were fixed into each strake as it crossed each frame timber and were placed diagonally across the join for maximum effect. In the nineteenth century it became common to use two metal bolts in the butt end of each plank, one above the other. This meant that four bolts were driven into the frame timber, because two ends joined together. In 1779, one bolt of 1in diameter was used in the butt end of each plank, clenched in the footwaling inboard. The spirketting of the gun deck and upper deck also had its butt end bolted. The hooding ends were bolted, because of the extra stress in that area.

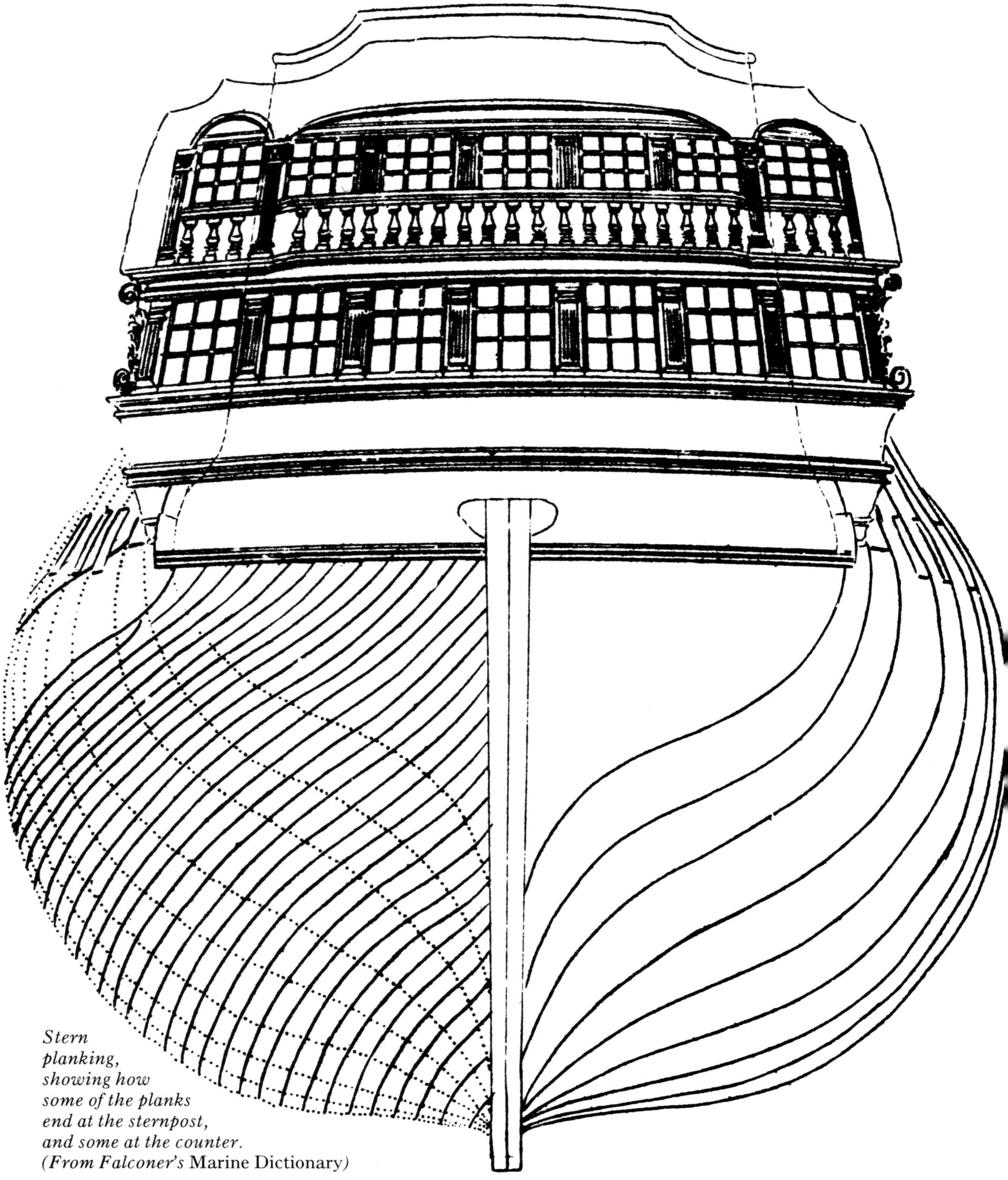

Stern planking, showing how some of the planks end at the sternpost, and some at the counter. (From Falconer's Marine Dictionary*)*

The Sides

The planking between the wales was generally easier to apply, because the sides of the hull had comparatively little curve. Of course, it was interrupted by the gun ports; in midships and forward, at least one strake passed under the ports without being cut, but the next strake up was cut with recesses to fit the shapes of the ports, and above that, short planks were needed to fit between the ports. Aft, the ports cut deep into the main wale, and the lowest plank was cut in short lengths; the plank above the ports was uninterrupted here.

Thicker planks were fitted just above the wales, similar to those below. The first strake above the wale, the 'thick stuff', was 6¾in thick, and the next one was 5¾in. However, there was no attempt to reduce the thickness gradually by tapering, as was done below the waterline, and there was a quite definite step in the timbers above the first strake of thick stuff. The rest of the plank above the main wale, up to the channel wale, was 4in thick.

Above the channel wale, the plank was 4in thick, tapering to 3in. Another band of thick planking, known as the 'sheer strake', almost serving as another wale, was fitted directly under the level

A model of an East Indiaman of the mid-eighteenth century, showing the underwater planking of the bow.
(Author's photograph, courtesy of the National Maritime Museum)

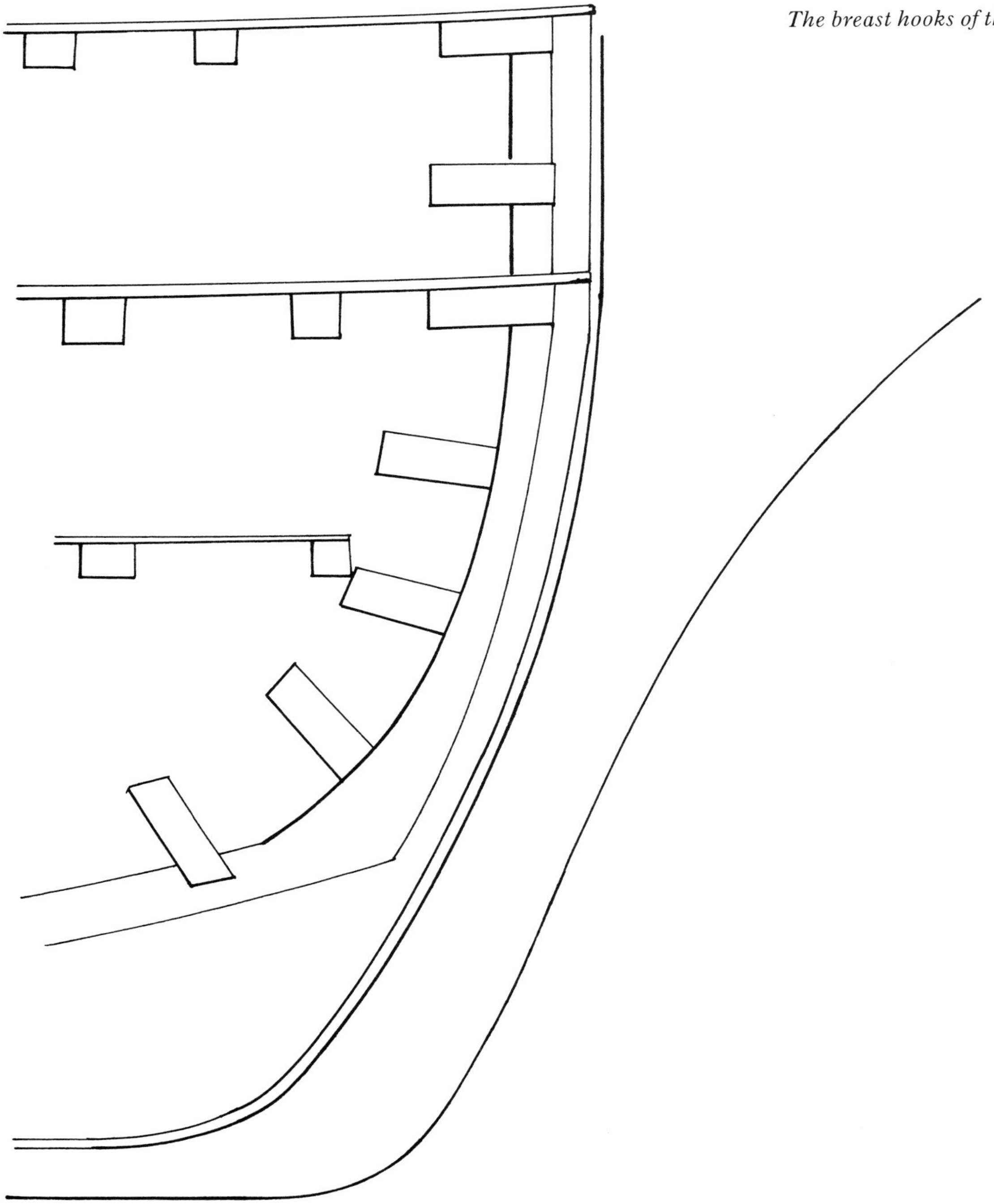

The breast hooks of the Valiant.

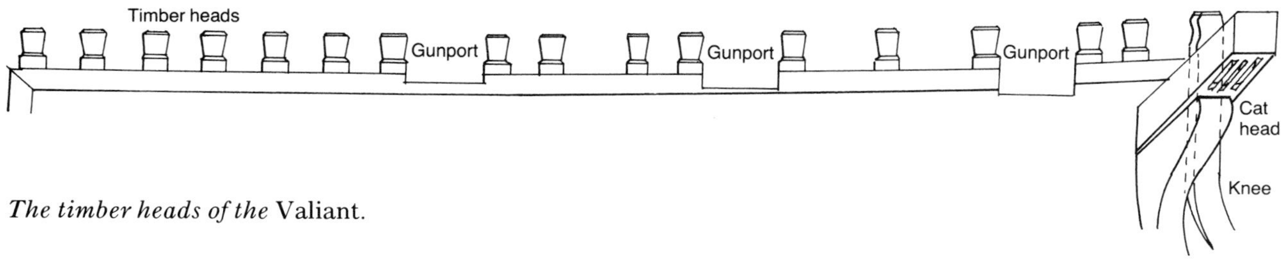

The timber heads of the Valiant.

of the top of the side in the waist, and it was said that 'it forms the chief strength of the upper part of the topside, and it therefore always worked thicker than the other strakes, and scarphed with hook and butt between the drifts'.[2]

The sides of the ship were finished off by a flat piece of timber on the gunwale, placed over the heads of the timbers and the top edges of the planking. Under the gunwale, in midships, was a rail known as the sheer rail. It was shaped in a decorative cross section, made from oak or elm, and nailed on top of the planking. It continued to the bow and the stern, following the line of the sheer. About 18in below it was a similar rail, known as the waist rail. At intervals above the sheer rail were yet more decorative rails, especially in the stern, where the structure was higher. The drift rail began just forward of the mainmast and the top of the side itself was raised at that point, by means of a decorative curve known as a drift. The drift rail began as the gunwale of a small section of hull, and then continued aft as a rail. Above the drift rail was the planksheer. This too began as a section of gunwale, but after the side rose up by means of a drift, it continued as a section of rail. The highest part of the side of the ship was at the poop deck; this too had a planksheer, which

Part of the internal planking of the Arrogant *of 1761, showing the use of clamp, spirketting and of angled knees to avoid the gunports.*

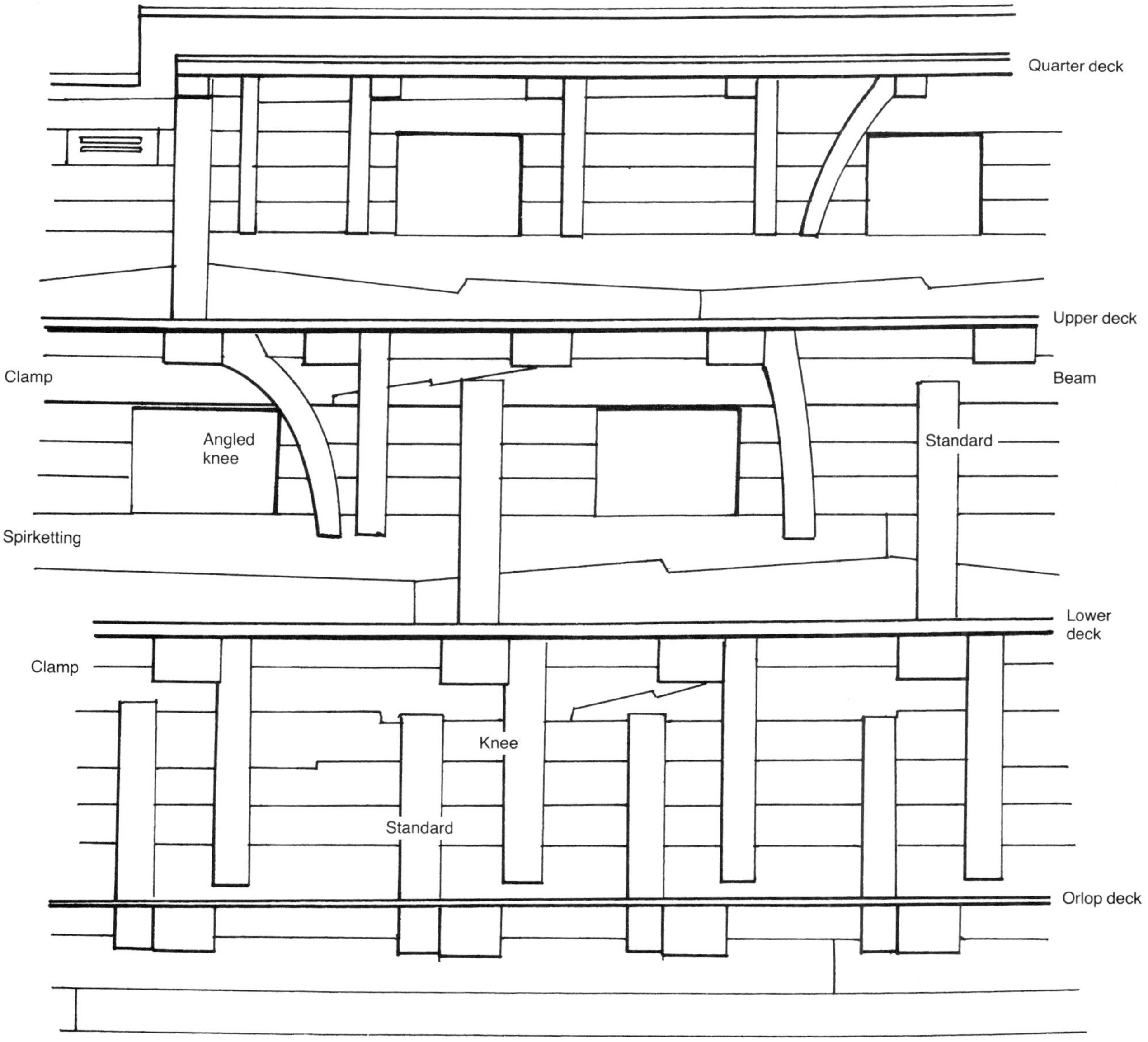

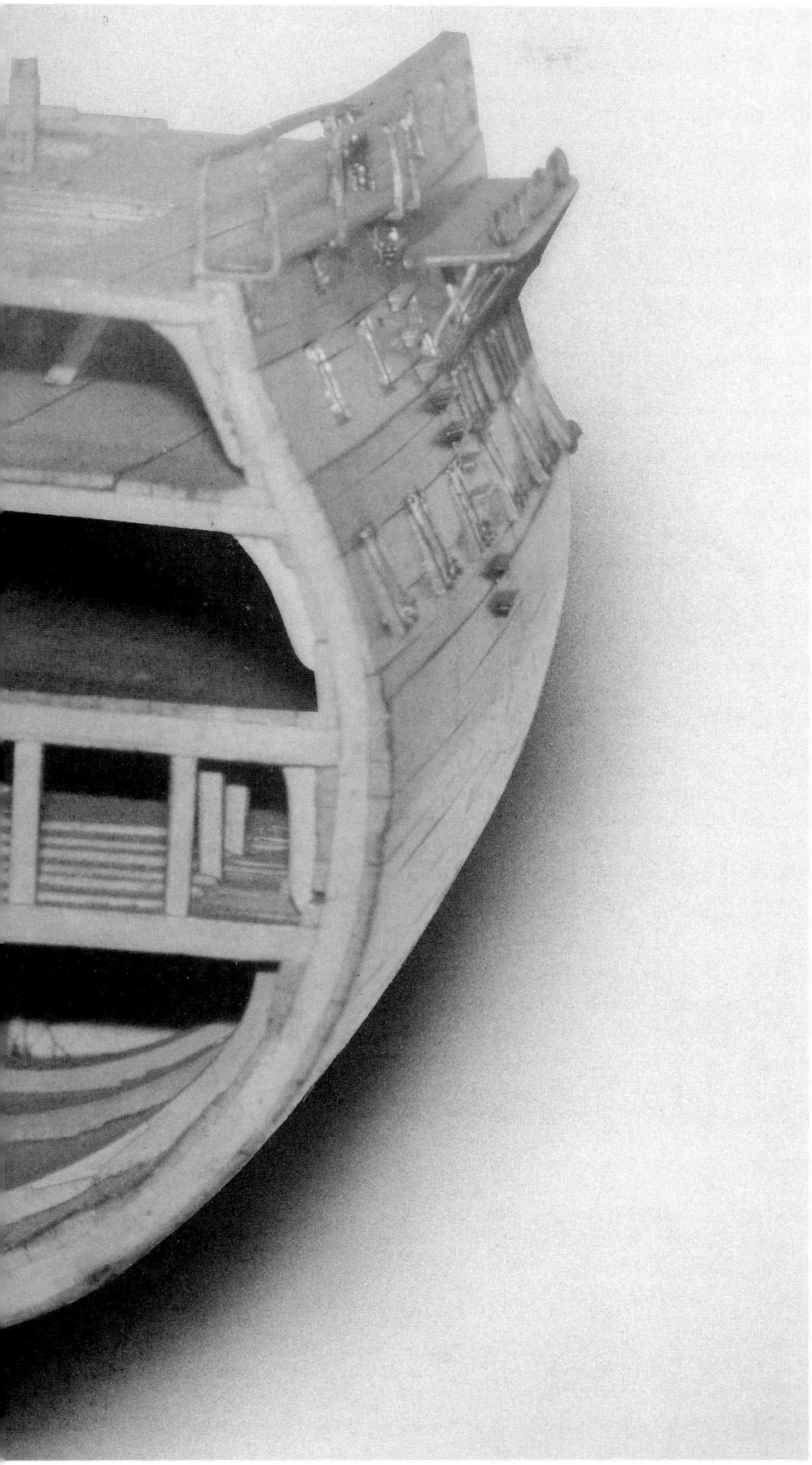

The midship section of a 74-gun ship, showing the riders in the hold and the pillars supporting the decks.
(Author's photograph, courtesy of the Science Museum)

served as a gunwale in that part.

Forward of the waist, the structure was rather different. The heads of some of the frame timbers were carried upwards above the level of the planking, and then bevelled to serve as belaying points for the rigging. These were known as the timber heads. The planksheer, therefore, had to be cut to fit round these heads.

Light vertical pieces, known as skids, fitted on the outside of the planking to protect the sides against objects, especially boats and casks, which might be hauled up the sides. Each skid was a piece of oak, cut to the cross section of the ship, and nailed on the ouside of the plank. One pair were fitted close together, in the after part of the waist, and were used for 'parbuckling' casks up the side. A single skid was fitted well forward, near the foremost part of the waist. In combination with one of the other two, it helped keep a boat off the side when it was being hoisted.

Internal Planking

The hull was also planked internally. The 'clamps' were probably fitted first. These served to support the deck beams, and had recesses cut in them to receive each beam. Naturally the scantlings of the clamp varied from deck to deck, but it always followed the line of the decks. The lower deck was of course the most important; according to a plan of 1771 and the contract of 1779, it was to be made in three strakes, with the pieces scarphed together with hook and butt. The upper strake was to be 8¼in thick and 1ft 5in broad, and the middle to be 7¼in by 1ft 2in, and the lowest was to be 6¼in thick. Since the deck beams were not fitted directly to the frame timbers, the clamps formed an essential part of the structure because, along with the knees, they bore most of the weight of the decks and their guns.

The internal planking of the hold was known as the 'footwaling' or 'ceiling'. It too had its thick stuff, placed over the joins of the futtocks of the frame, in order to strengthen them. The first band of thick stuff was placed near the keel, covering the chocks of the first futtocks. The main strake, 7½in thick, was placed a few inches away from the keelson. (The resulting hole was used to drain the bilge water towards the pumps, and was later covered by a row of planks placed diagonally between the upper corner of the strake of thick stuff and the keelson.) There were two strakes of thick stuff over the chocks of the first futtocks, 7½in and 6½in thick. As with the rest of the thick stuff in the hold, they were not faired in with the rest of the plank, but the corners were usually trimmed off to give a slightly smoother appearance. The next band of thick stuff was over the heads of the floor timbers. Five strakes were used here, three in the middle, 8¼in thick, and one on each side, 6¾in thick. This band tended to rise quite sharply, and merged with the clamp of the orlop deck before it reached the bow and stern. A rather lighter band of thick stuff was fitted over the first futtock heads – one strake of 7in, and one on each side of 6in. No further bands of thick stuff were used; the second futtock heads in midships were at almost the same level as the clamp of the orlop deck. The spaces between the thick stuff were filled in with 4in plank, following the run of the thick stuff.

The rest of the internal planking followed the lines of the decks rather than the futtock heads, and therefore rose much less steeply. It was generally fitted after the deck beams were in place, for otherwise it would have been difficult to fit the beams into the narrow spaces between the planks. The three tiers of clamps which supported the lower deck took up most of the space between that deck and the orlop, leaving room for two other strakes. A 6in plank was placed just above the beams, angled to join with the waterway, which was part of the deck planking. A single strake of 4in plank was fitted between this and the lowest strake of clamp. The spaces along the side between the orlop deck beams were filled with 6¼in plank.

The space between the lower and the upper deck was similar, though the clamp was less extensive – only one strake was used, 5¼in thick at the top edge and 4¼in at the lower. The next two strakes of ceiling, on a level with the gunports, were of 3in plank. Below that, between the lower edges of the ports and the waterways of the gundeck, were two strakes of spirketting, 6in thick. This was regarded as an important contribution to the strength of the ship, and special care was taken with the selection and seasoning of the timber, and with the positioning of the butt ends to give maximum strength.

The use of clamp and spirketting continued under the quarterdeck and forecastle, though with thinner pieces of timber. Another strong piece, known as the 'string', was placed just under the gunwale in the waist, in order to strengthen that area, where the support of the upper works of forecastle and quarterdeck was not available. The string consisted of two strakes of 4in plank, scarphed into the ends of the forecastle and quarterdeck clamp.

Internal Bracing

In addition to the timbers of the frame, the hull had a system of bracing inside the internal planking. In midships, this consisted of floor riders, first futtock riders and second futtock riders, constructed in a similar way to the frame timbers, though much less numerous. In midships, where the floor was at its flattest, a floor rider was shaped to fit over the keelson extending about 14ft on each side over the planking. To its ends were joined second futtock riders, reaching up to the lower edges of the lower deck beams, passing through the orlop. Alongside these were the first futtock riders, beginning within a few feet of the keelson, and reaching up to the orlop deck beams. The riders were securely bolted to one another, and also to the timbers of the frame. The full system was used for four sets of riders, extending from just aft of the mainmast to the middle of the waist. Forward and aft of that the timbers tended to rise more steeply from the keel, and the system was modified so that the first futtock riders were further from the keelson. Thus the system of riders continued for about half the length of the hull. There was some attempt to space them so that one was placed between each pair of gunports on the lower deck, but this had to be modified in practice, so that the foremost pair helped to support the capstan partners.

The mast steps were similar to floor riders in form, in that they passed over the keelson and the internal planking. Each provided a seat for the lower end of a mast, and was cut with a square hole to receive it. The foremast step was made in three parts, being 11in square in cross section and 24ft long over the keelson. The mizzen step was 1ft 8in square, and 14ft long. The main step was 2ft 11½in by 1ft 3in, but was considerably shorter than the others, because it had to fit inside the well of the pump.

In the bow, a system of 'breast hooks' was used for internal strengthening. These were put in the space between the foremast step and the lower deck and at right angles to the stempost. The highest one, set in the horizontal plane, was known as the deck hook for it formed a support for the foremost part of the upper deck. The third one was also a deck hook, and performed a similar function for the lower deck. In all there were seven deck and breast hooks, approximately 3ft by 14in in cross section, and about 19ft long over the stempost.

At the stern there were further variations. Two 'crutches' were placed over the deadwood between the mizzen mast and the stempost. These were V-shaped, about 8ft high and angled a little from the vertical. They were some way forward of the transoms, and not directly connected with them. The transoms had a separate system of bracing in the horizontal plane. At the level of the gundeck, and forming part of its structure, were two transom knees angled between the deck transom and the side of the ship. In 1779 the arm attached to the transom was 6ft long, and the other was 16ft long.

10

The Decks

A 74-gun ship like the *Valiant* was known as a two-decker, meaning that she had two complete decks of guns. In fact, she had decks of some kind at six different levels and carried guns at three of these levels (four levels in later years, after ships were fitted with carronades on the poop). Each deck served different functions and was constructed accordingly. All formed a vital part of the structure, in that the heavy cross beams which supported them helped brace the hull against the pressure of the water.

The Deck Levels

In many ways the most important deck was the lower deck, fitted a few feet above the waterline. It was also known as the gundeck, the lower gundeck, or the main gundeck because it carried the heaviest of the ship's guns: 32-pounders in the case of the *Valiant*. Because of this it had the strongest beams and they were the most important in bracing the hull, especially since they were situated close to the waterline where the stresses were greatest. The lower deck also served as the eating and sleeping accommodation for the crew, and as a working area for taking in the cables when raising anchor.

The next deck up was the upper

The longitudinal section of the Valiant, *showing the deck layout.*

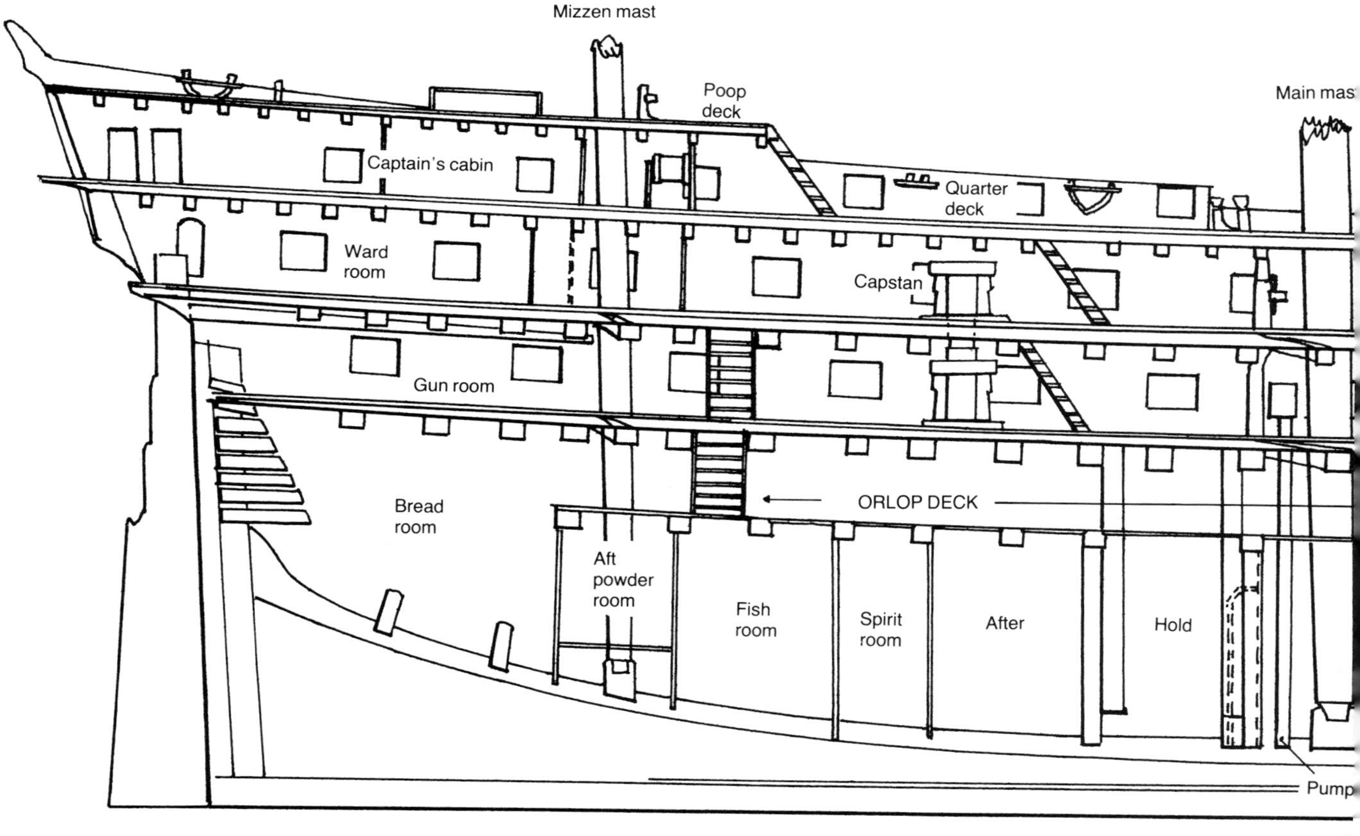

deck, or the upper gundeck, and it was sometimes known as the main deck because it was the only complete deck which was exposed to the weather in any part. It too had a full range of guns, 24-pounders on the *Valiant* and 18-pounders on the great majority of 74-gun ships (and on the *Valiant* herself after the 24-pounders were found to be too heavy). The after part of the upper deck (amounting to almost half the length of the ship) was covered by the quarterdeck. About half the enclosed space thus created (that part nearest the stern) served as the wardroom for the officers while the rest of the space provided some measure of shelter for the watch on duty in bad weather, though it was open to the elements at its forward end. The foremost part of the upper deck was covered over by the forecastle, and contained the galley, and perhaps some accommodation.

The quarterdeck carried fourteen 9-pounder guns. It held the steering wheel and the compass binnacle, so it was the main area from which the ship was controlled and commanded. Its after half was covered by the poop deck, and under that was the captain's cabin, a small office for his clerk, and a chart room for the master, next to the steering wheel. It had belaying points for many of the ropes of the standing rigging and was the main recreation area for the officers.

The forecastle was fitted in the bow, at the same level as the quarterdeck, and covered about a quarter of the length of the ship. It was armed with only four 9-pounder guns but it served as a base for much of the rigging of the foremast and the bowsprit. It was also used for access to the catheads, which were used in hauling up the anchor, and as a platform for lookouts in confined waters.

The space between the quarterdeck and the forecastle was known as the waist and until about 1740 it had been completely open. After that gangways were fitted between the quarterdeck and forecastle, running along the sides of the ship, and, over the years, they tended to become wider and more permanent. Booms were crossed between them, to hold the ship's boats, and little light could have reached down to the upper deck. Indeed, on *Victory* today the space is also covered with canvas and it is difficult for the casual visitor to appreciate that there is not a complete deck above the upper deck.

The poop deck was in the stern, above the quarterdeck, and took up about half its length. Apart from covering the captain's cabin and the steering wheel, it served as

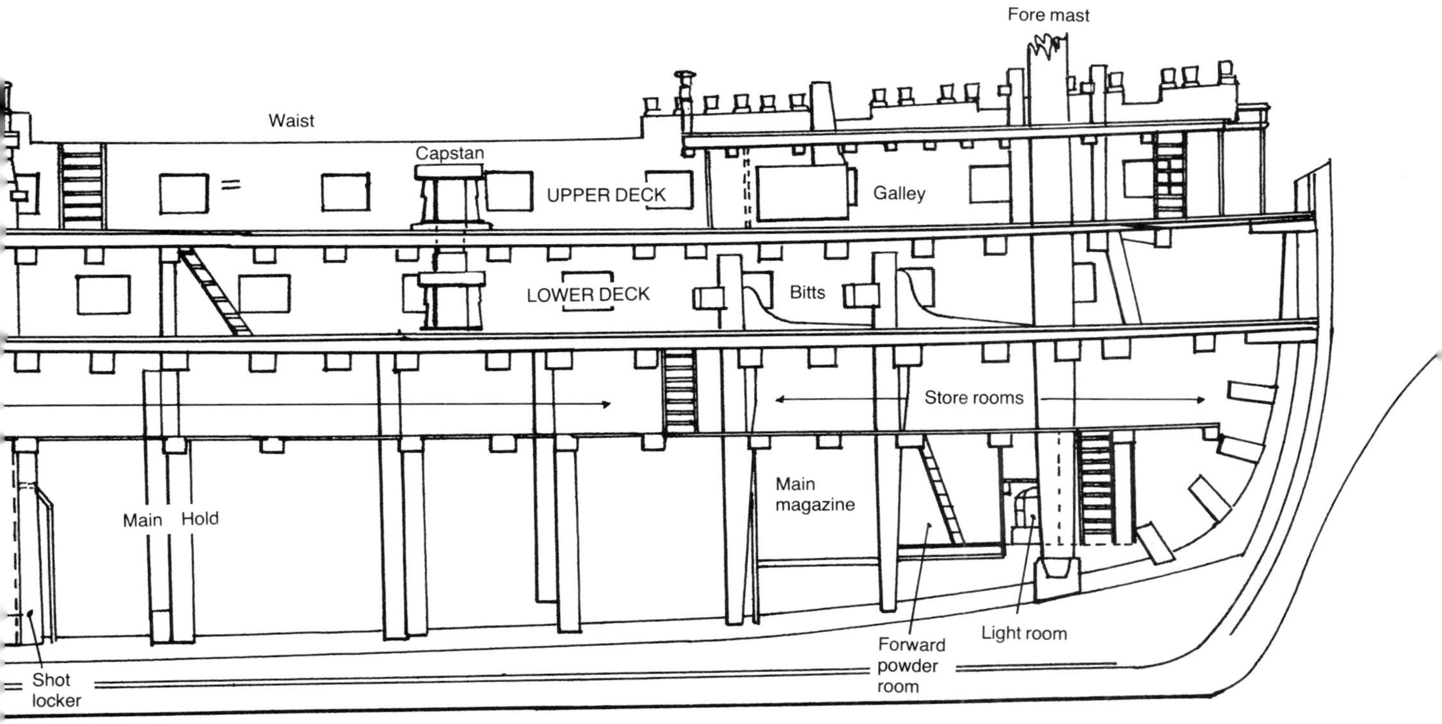

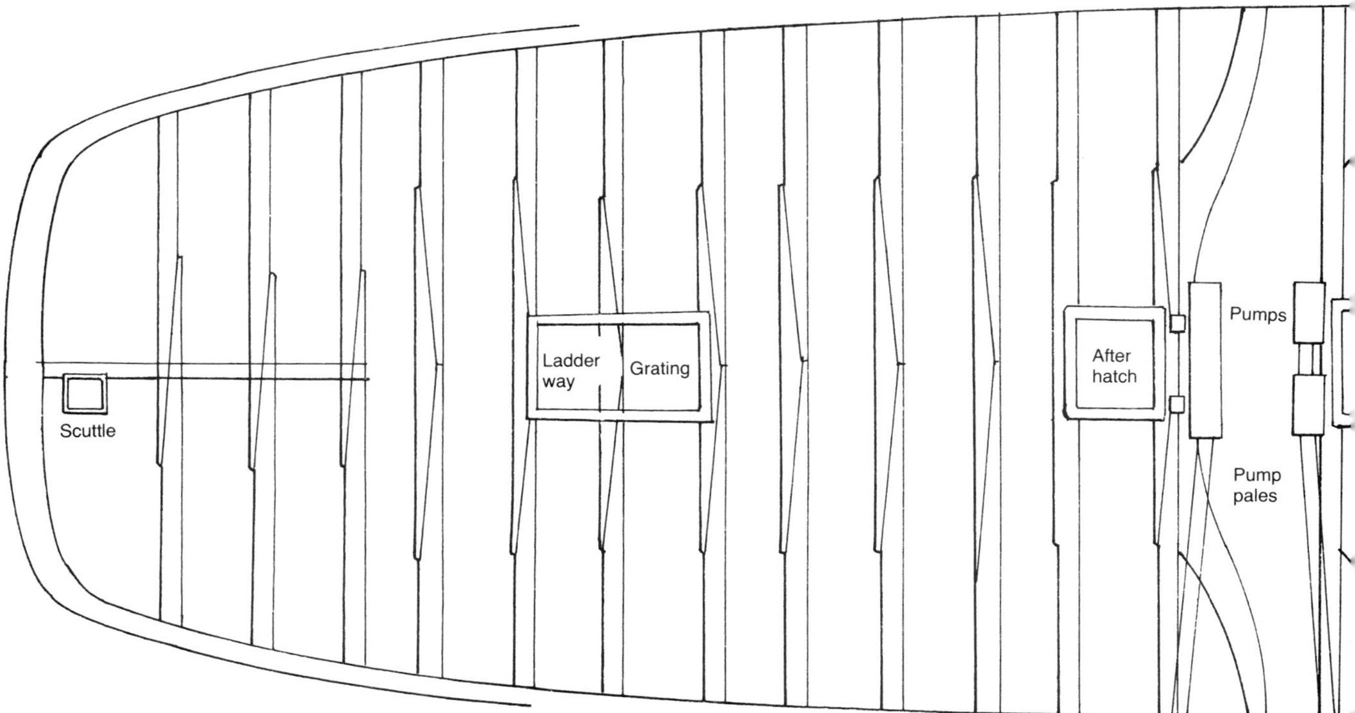

The lower deck of the Valiant. *The carlines and ledges are not shown.*

The upper deck. The carlines are shown on one side, but not the ledges.

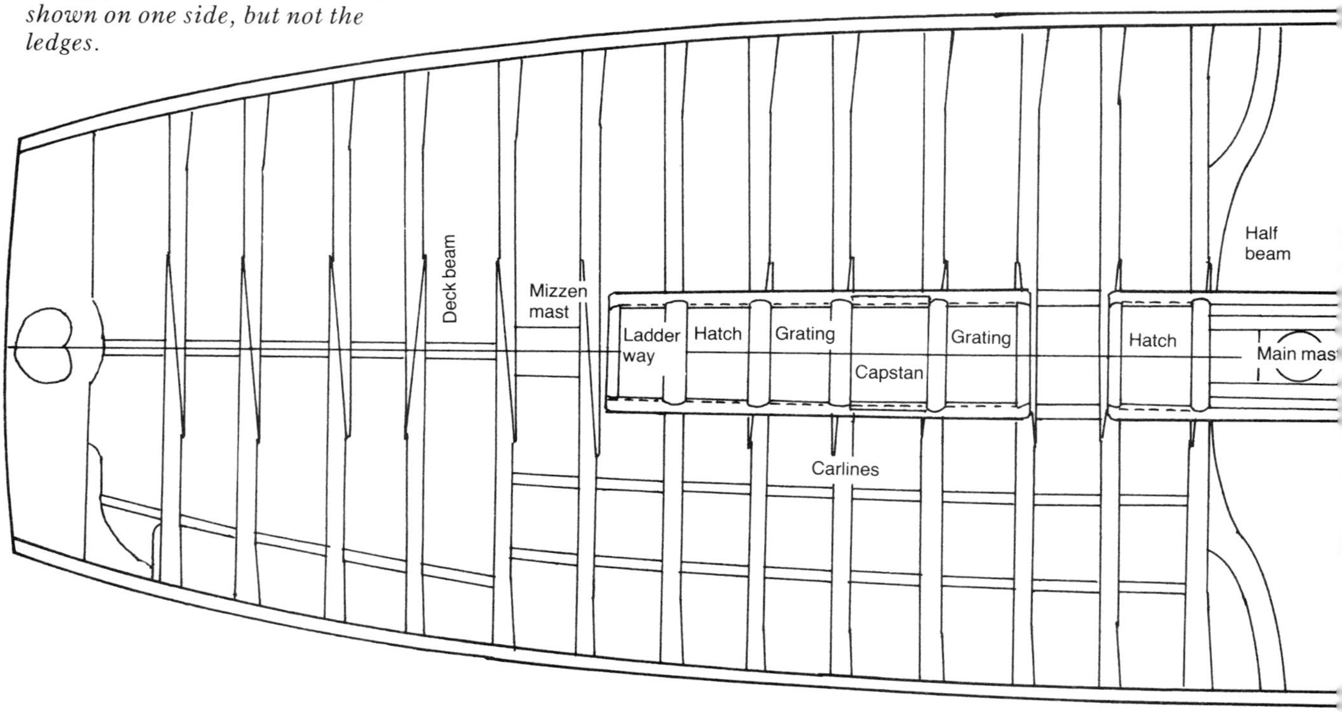

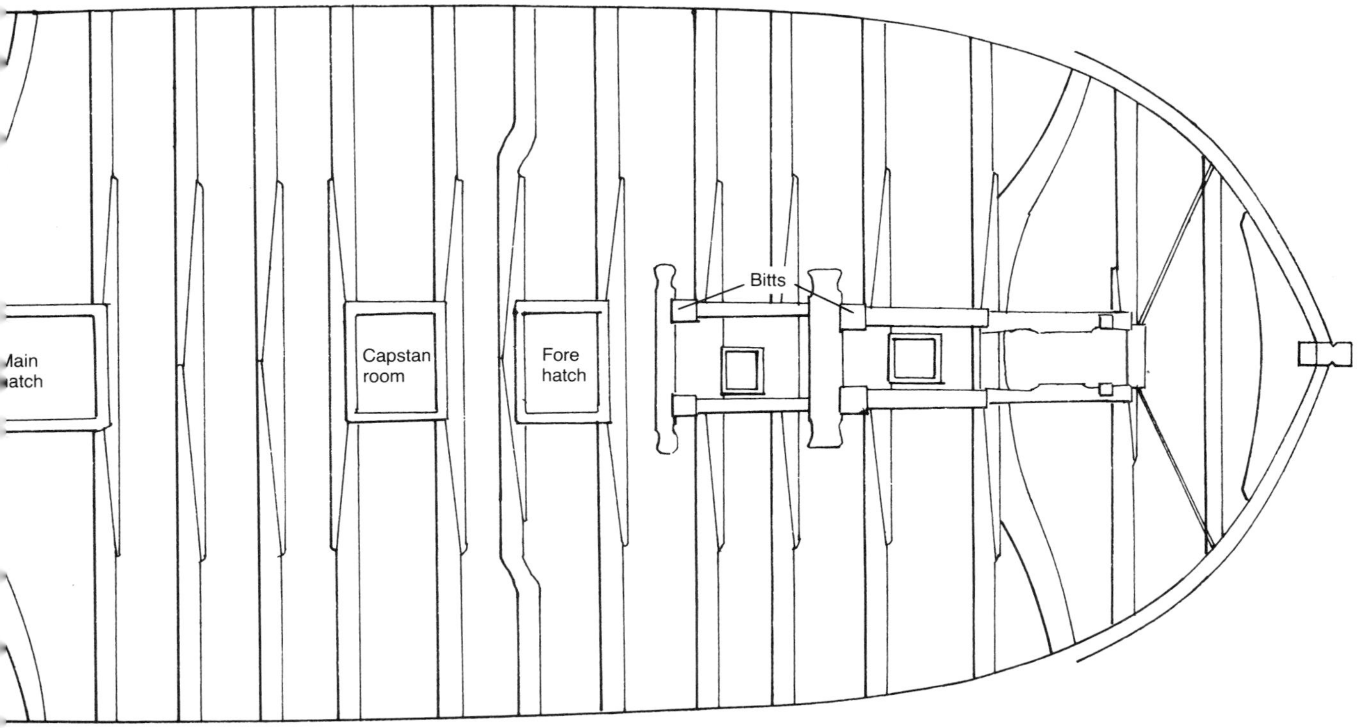
Bitts
Main hatch
Capstan room
Fore hatch

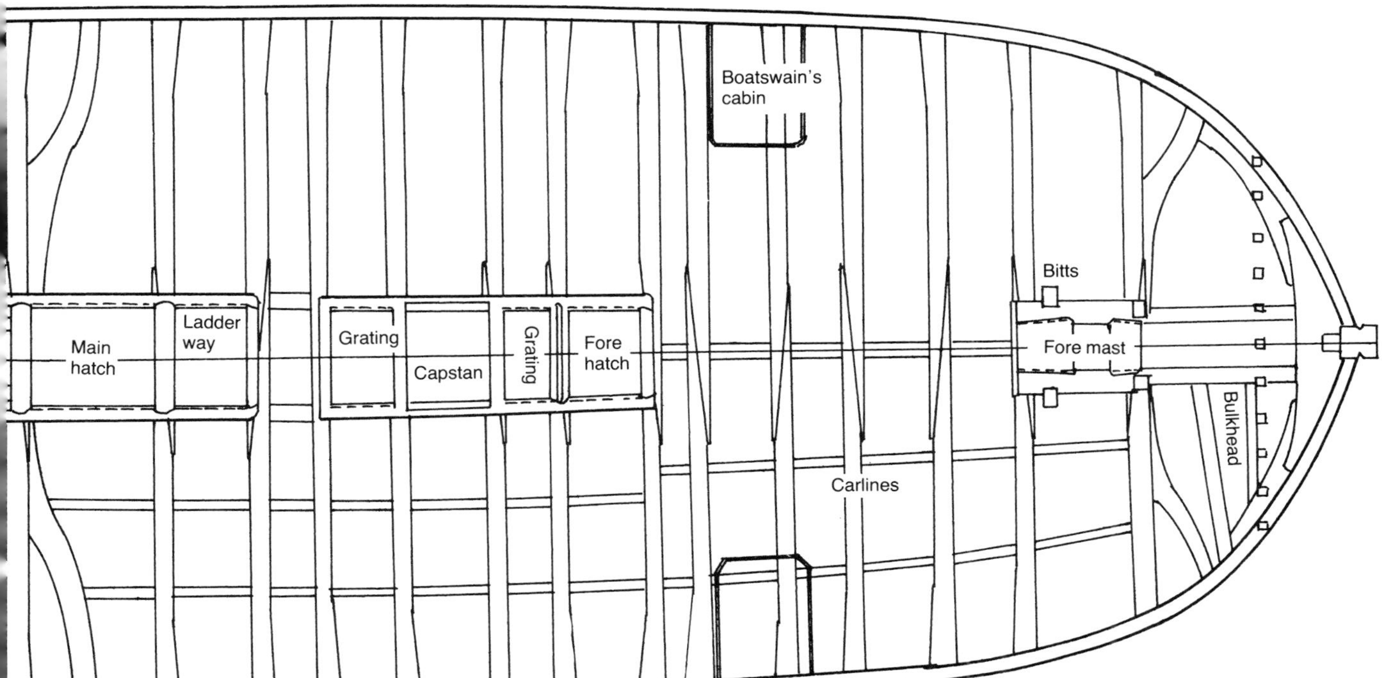
Boatswain's cabin
Bitts
Ladder way
Main hatch
Grating
Capstan
Grating
Fore hatch
Fore mast
Bulkhead
Carlines
Carpenter's cabin

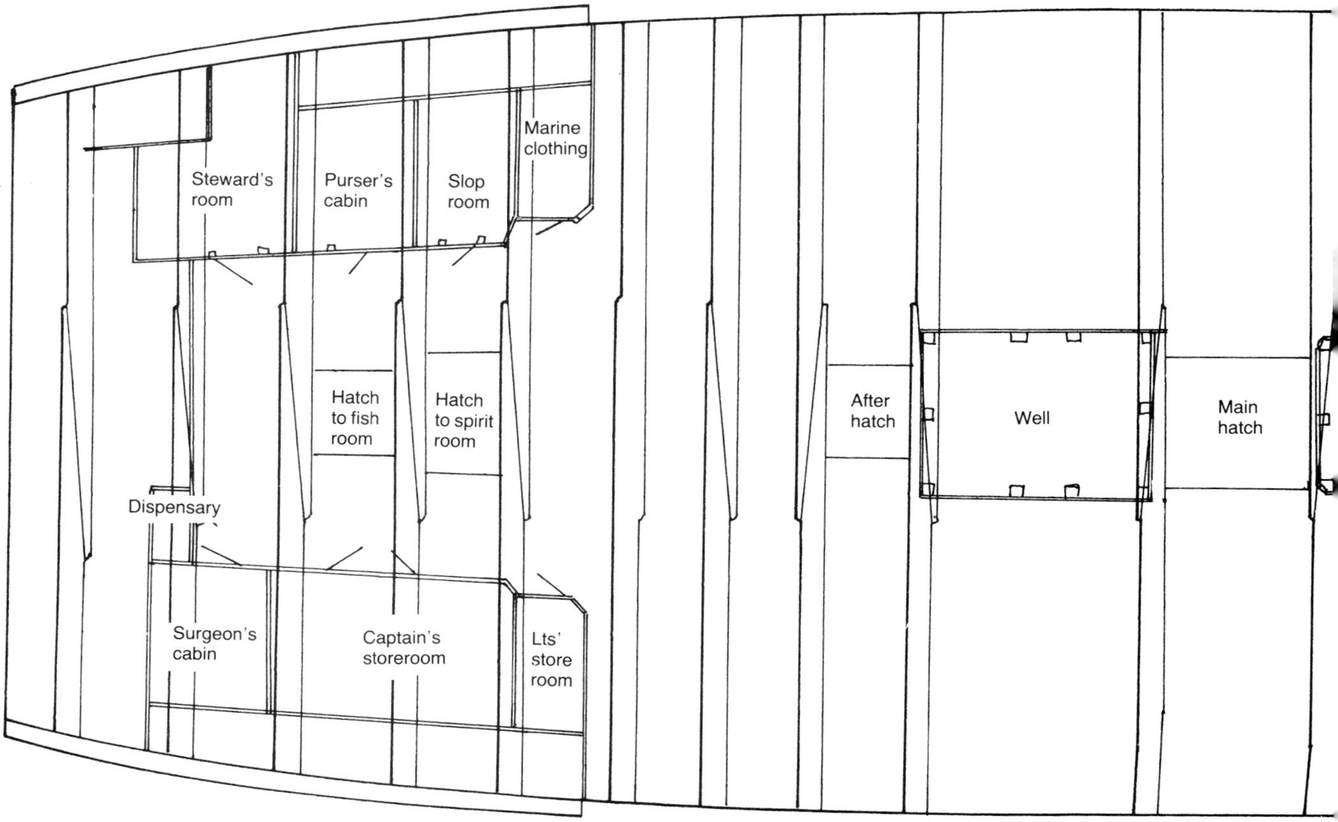

The orlop deck showing the division into cabins and store-rooms.

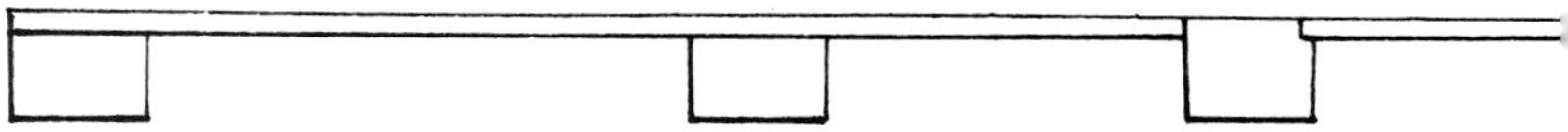

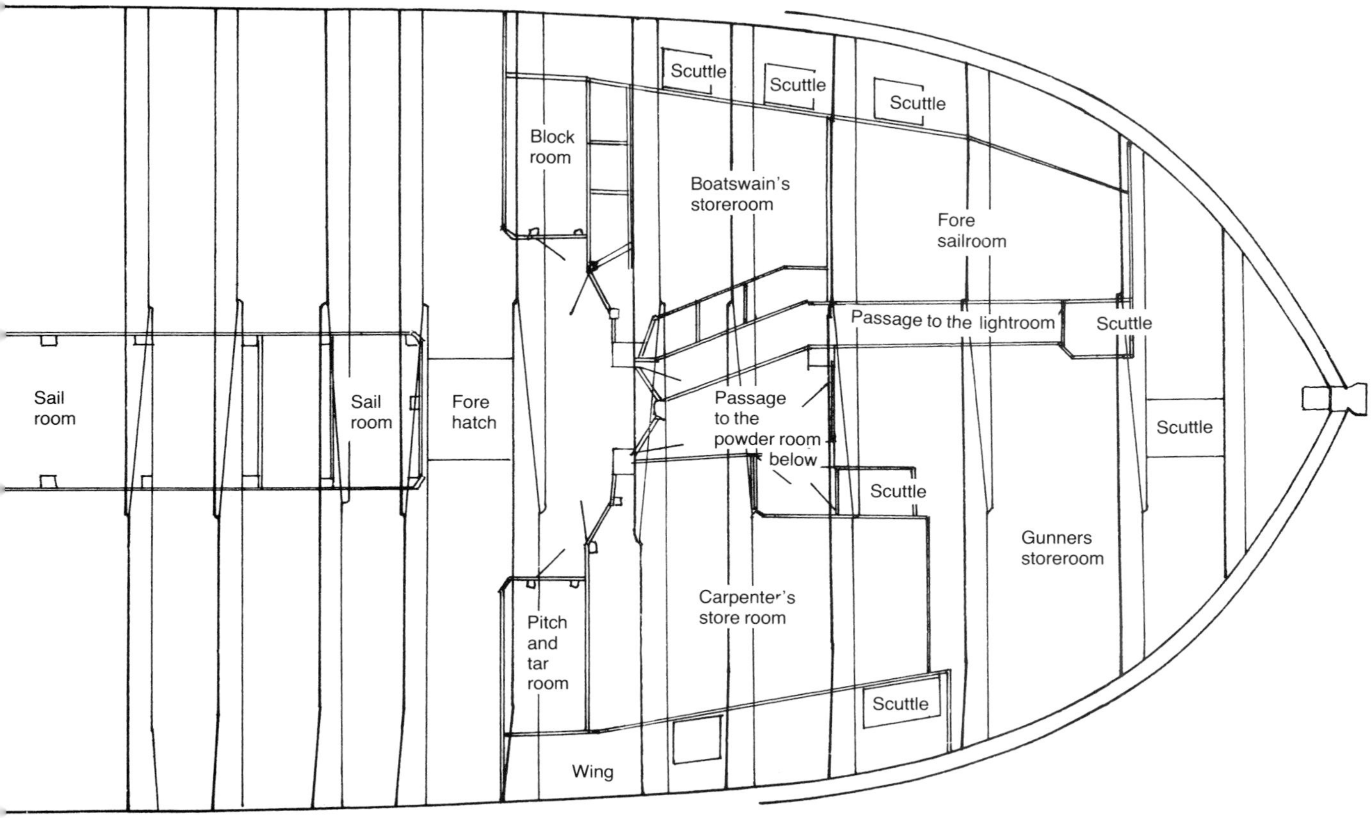

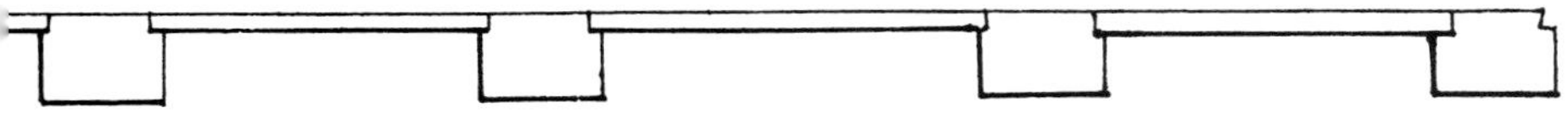

The planking of the orlop deck, with a section of more conventional planking to the left.

An old deck beam from the masthouses at Chatham with the original scarph still visible.

a useful vantage point for conning the ship, and as the area where much of the running rigging, especially that of the mizzen mast, was handled. It carried no guns until after 1779 and then it became common to fit carronades there.

The orlop deck was the only complete deck below the lower deck. It was slightly below the waterline and, of course, carried no guns. Its construction differed from the other decks, in that its deck beams were not cambered, and its planks in midships were much shorter, being fitted between the pairs of beams rather than across them. The orlop deck provided more strength for the hull below the waterline, and it had several important store rooms – for the purser and surgeon aft, and for the boatswain, carpenter and gunner forward. It contained cabins and other accommodation for warrant officers and midshipmen, and sail rooms where spare sails were kept. A large area in midships was taken up by the anchor cables, which were stored there in enormous coils. The short planks were useful there, because they could be taken up to allow the water from the cabins to drain into the hold.

The hold was mostly undecked except in the area of magazines and powder rooms. These structures did not form an integral part of the ship like the decks, and therefore can be treated as fittings rather than as part of the structure.

Deck Beams

The deck beams were the most important parts of the structure of the decks and they were heavily built. The lower deck beams of the *Valiant* were 1ft 4½in square in cross section – while even her poop deck beams, not intended to carry

The structure of an upper deck from a model of the 1790s. It is more realistic than most models in that it shows the carlines tenoned into the beams, etc. However, the capstans and gratings would not be fitted before the deck planking. (Author's photograph, courtesy of the National Maritime Museum)

any great weight, were 7in by 5½in. Not only did they support the decks and guns but they helped brace the ship against lateral pressures. Before the ship was launched, or while she was in dry dock, they prevented the frame from falling outwards and distorting the shape. When the ship was in the water, they braced the hull against the great pressures of the sea. The beams of the orlop deck, though they bore no guns, were nevertheless 15⅝in square on a 74-gun ship, because this deck, being closest to the waterline, had an enormous amount of weight.

All the deck beams, except those of the orlop, were cambered with their centres higher than their outside edges. This allowed water to drain to the sides and it may have helped to restrain the recoil of the guns and make it slightly easier to run them out. The orlop was below the waterline and, of course, carried no guns, so was built flat. The camber tended to increase with the higher decks so that the gundeck beams of a 74 rounded up 5in, the forecastle 7in and the poop 11in.

The deck beams of a ship of the line were generally too long to be made from a single piece of timber and in most cases three were scarphed and bolted together, with a long triangular piece forming the centre. Naturally the beams nearer the bow and stern were shorter, as were those of quarterdeck and poop, because of the tumble-home of the ship and so sometimes they could be made in a single piece.

The deck beams were not fixed directly to the frames of the hull. Instead, a thick plank known as the 'clamp' was fitted to the frame. It had small recesses to receive the deck beams, and was fitted at an appropriate level. Unlike ordinary planking, its ends were scarphed together. There were usually two or three strakes of clamp, of decreasing thickness. The contract of 1779 specified three, with thicknesses of 8¼in, 7¼in and 5in.

Often the spacing of the deck beams was interrupted by the numerous features placed along the centreline of the ship such as masts, capstans and hatchways. Rather than miss out a beam com-

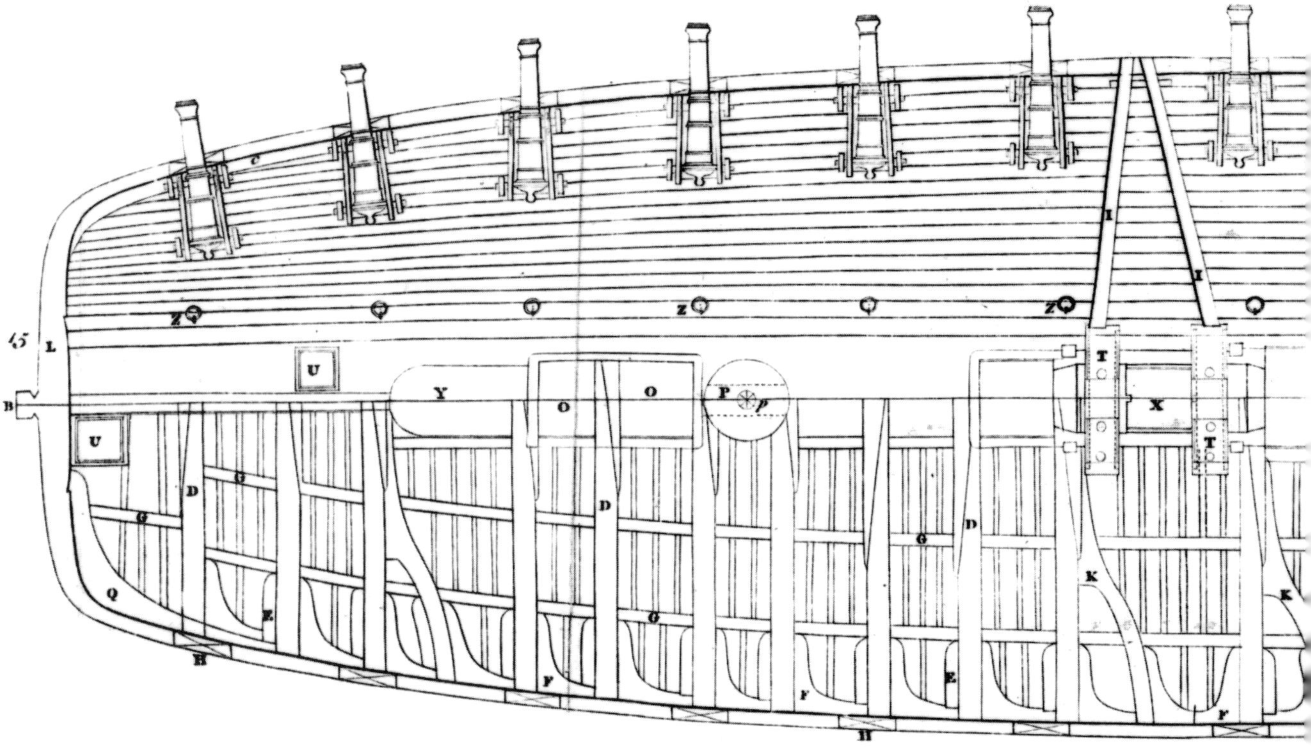
45
L
B
C
Z
U
Y
O
P
X
T
I
D
G
Q
E
F
H
K

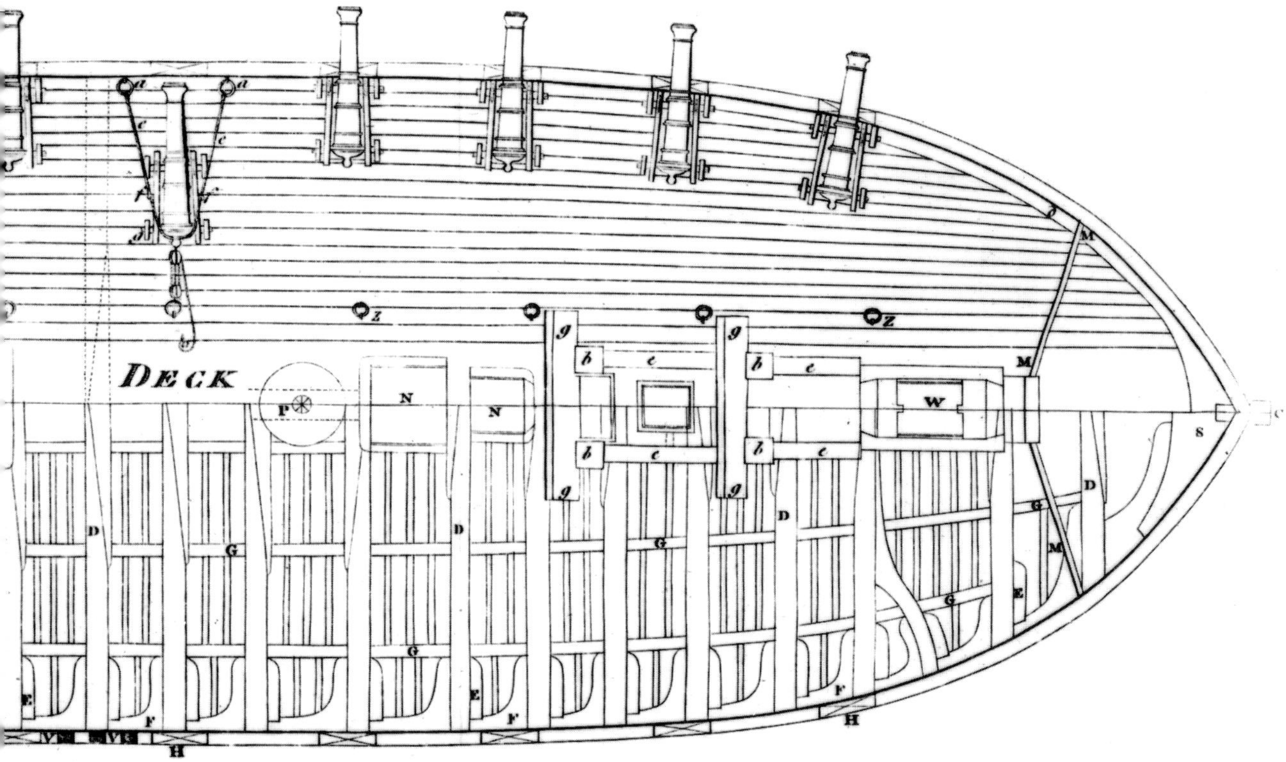

The deck plan of a 74-gun ship, showing the supportive structure and planking. (From Falconer's Marine Dictionary*)*

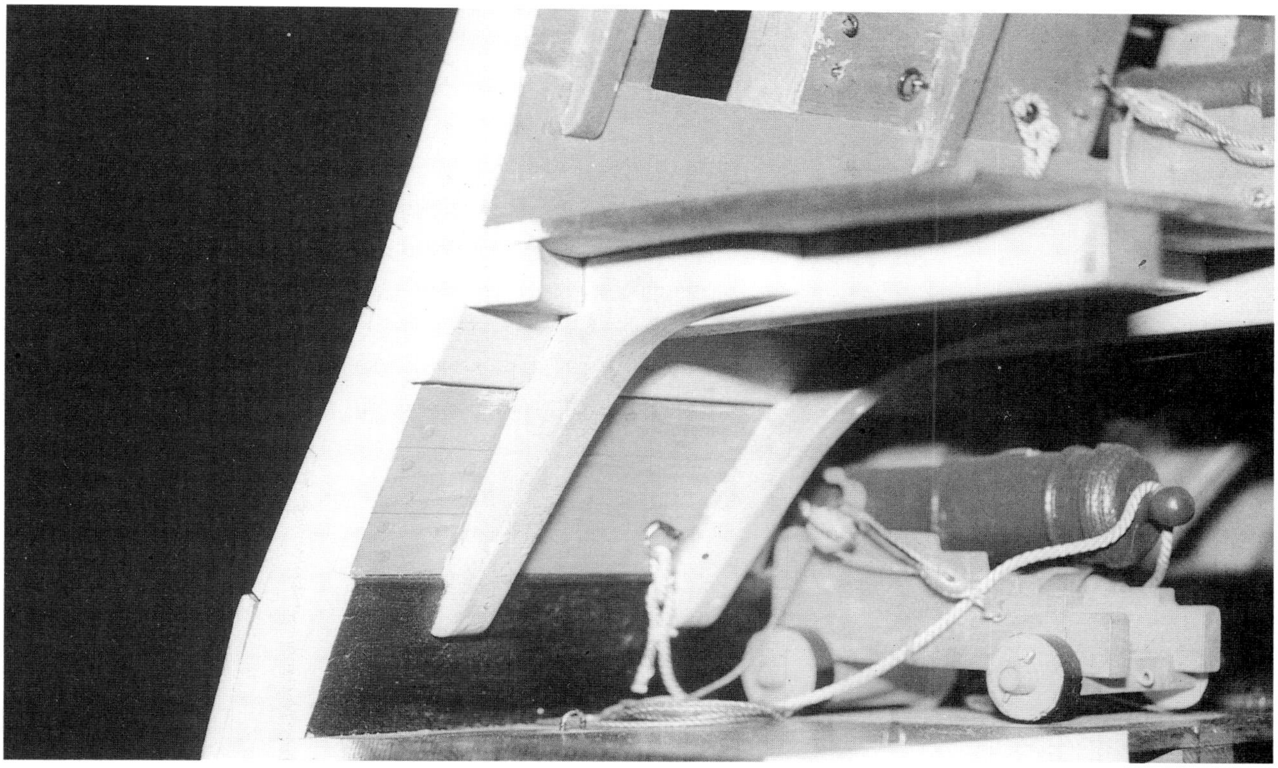

pletely to allow space for the main hatchway, for example, the shipwright would use half beams. These started out at the side of the ship in the usual way, but curved either forward or aft to join the adjacent full deck beam. As a result, the space in the centre of the deck was left uninterrupted.

Carlines and Ledges

The deck beams alone were not enough to support the plank of the decks against the weight of the guns. The beams were normally 4ft or 5ft apart, but the guns could be moved anywhere on the decks, and without additional support they would be held up only by the 4in thick plank of the deck. Two systems of bracing were used to support the plank between the deck beams: carlines ran fore and aft, and ledges ran athwartships, parallel to the deck beams. The carlines were approximately 9in square on the gundeck of a 74-gun ship and were arranged in three rows on each side of the central part of the ship, with one line forming the support of the outside edges of the hatchways, partners and other features on the centreline. The other two rows were evenly spaced between the first row and the sides of the ship. Naturally, the lines of carlines curved inwards from midships, but close to the bow and stern they became too close together. After that point, only two rows were fitted. The carlines were fitted into the deck beams at each end to help hold the beams together. Extra strong carlines were fitted in positions where they might bear strain, for example under the galley stove, forward on the upper deck.

The ledges were considerably smaller than the carlines, 5in or 5½in in section. They ran athwartships between the carlines and were dovetailed into them. Others, fitted between the outer tier of carlines and the side of the ship, were attached to the lodging knees. The contract of 1779 specified that the ledges be carefully spaced so that they were between 9in and 12in apart. This often led to irregular spacing, especially in the area of the lodging knees.

◄ *An old ship's knee in the masthouses at Chatham showing the direction of the grain.*

A part model of a ship of the 1750s showing the use of the hanging knee.
(Author's photograph, courtesy of the National Maritime Museum)

Hatches and Partners

The innermost row of carlines was less regular than the other two, because it formed the outside edge of the various features along the centreline of the deck. On the *Valiant's* gundeck these began about 8ft from the stem and consisted

The bitts and their fittings.

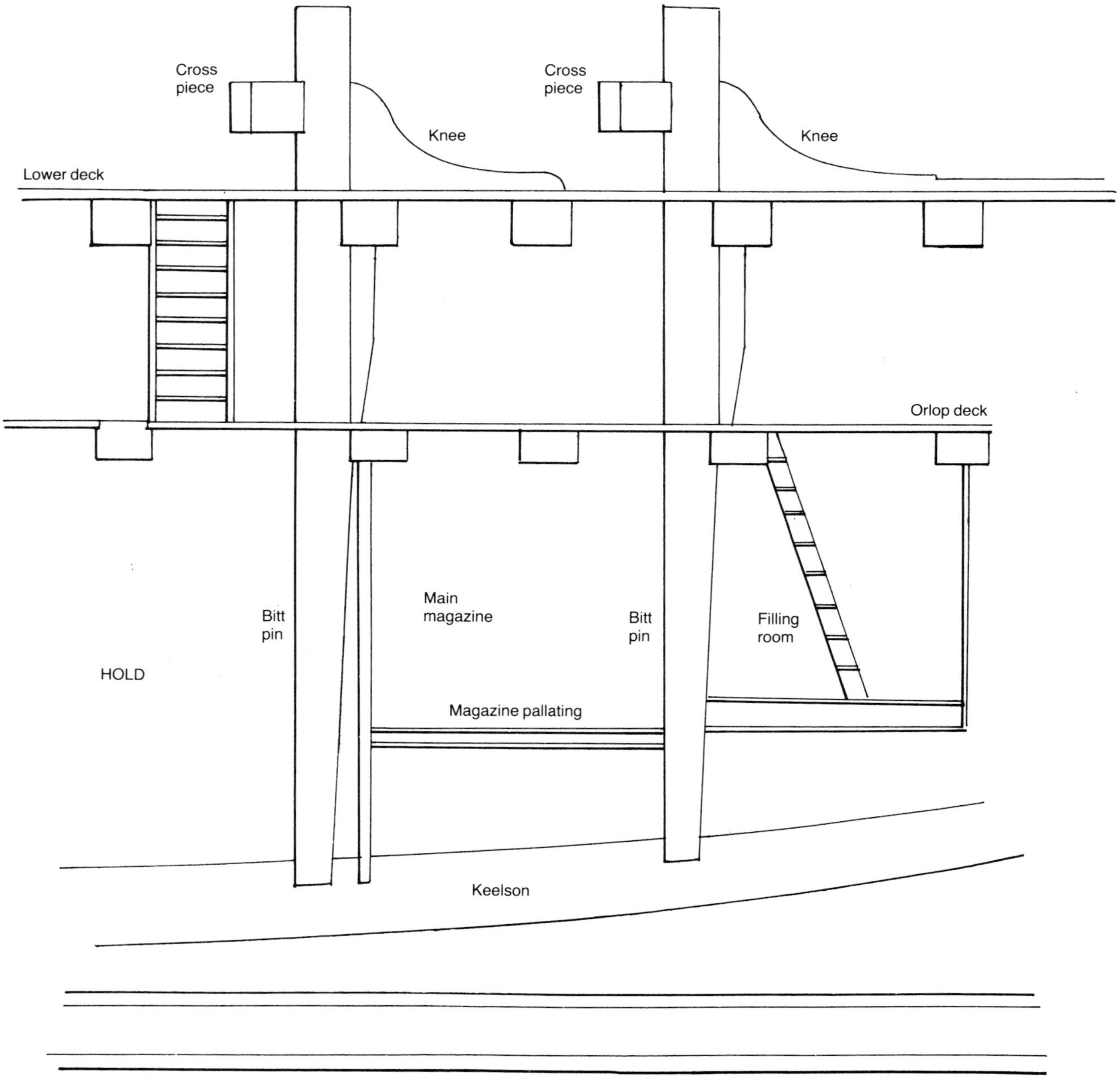

A cathead on a model showing the cat block rigged under it. (Author's photograph, courtesy of the National Maritime Museum)

of, in order, the partners (supports) of the foremast; two small hatches to pass ammunition from the magazine; a ladderway to the deck below; the fore hatch; a small storeroom below the fore capstan; the main hatch; the partners of the main mast, with the pumps spaced around it; the after hatch; a space below the main capstan; a grating to the fish room in the hold; a ladderway; the partners of the mizzen; and a small hatchway (or scuttle) right aft, to the bread room. The row of fittings was rather less continuous than it would become in later years, and there was a substantial gap between, for example, the capstan room and the main hatch. The positions were largely repeated on the upper deck, though the small scuttles were not fitted there, and the capstan room was replaced by the partners of the capstan which were much stronger.

The partners of the masts consisted essentially of holes in the decks through which the masts could be fitted. Each was a square piece of timber (9in thick in the case of the mainmast partners of the orlop deck) with a round hole in the middle, slightly larger than the mast. Partners were fitted between specially-placed carlines in the deck, with 'head ledges' to receive the forward and after ends. Capstan partners were different and were made in two halves so that they could be fitted under the whelps of the capstan, to fit with its spindle and allow it to turn easily without slipping out of place.

The hatches and ladderways had coamings fitted round them, above the carlines and ledges which defined their shapes. A coaming was simply a raised piece of timber which stopped water from running down into the hatch and deterred any seaman who might stumble up on it in the dark. Each was cut with a recess in its inner edges, in order to hold a grating or hatch cover.

The ends of the planking, at the bow and stern, were supported by deck holes and deck transoms respectively. These were part of the general structure, but were specially placed so that their top surfaces would form a solid basis for the ends of the decks. The transom was slightly different in principle from the hook; the former was part of the frame of the ship, while the hook was part of the internal bracing. At upper deck level the deck hook was quite short on the *Valiant*, but it was aided by a knee on each side which brought the structure round to meet the first half-beam.

Knees

Knees formed an essential part of the structure of the decks. Knees, in their traditional form and as used up to the end of the eighteenth century, were L-shaped pieces of timber and came in three basic types. Hanging knees were placed vertically and supported the deck beams from below; lodging knees were placed horizontally, in the angle between the beam and the side of the ship; and standards were similar in function to hanging knees, except that they were placed above the beams, rather than below. Practically every beam had a hanging and a lodging knee, but standards were much rarer; a 74 had only twelve standards on each side of her gundeck, compared with twenty-seven hanging knees. Knees could vary considerably in shape according to their position in the hull. Where the tumble-home was slight, their arms would meet at right angles; where the tumble-home was greater, the angle would be obtuse. Where a knee coincided with a gunport, it was often necessary to angle it from the vertical, or even to make it with a pronounced curve in the fore and aft plane.

The purpose of the hanging knee was not merely to support the deck against downward pressure but it also acted against the twisting motion of the hull in the sea. Both arms of the L needed to be quite long so that one could be bolted securely to the deck beam and the other to the side of the ship. The horizontal arm was placed alongside the beam, not under it, and this saved some space near the sides of the ship; as the top side of the knee was pushed right up against the deck it also helped to support the weight of the guns.

In the 1750s warship knees were generally made of wood. The principle of iron knees was known in both Britain and France, and indeed the *Invincible* had been fitted with such knees. They were remarked on by the officers of Portsmouth Yard who surveyed the ship in 1747, and they must have been considered effective since they were still on the ship when she was lost eleven years later. However, it seems likely that the *Valiant* was fitted with conventional wooden knees. In later years a kind of hybrid knee was evolved. The wooden part was basically a triangle, fitted under the deck beam, and a strong iron bracket was fitted over this so as to make a rigid join between knee and beam. Such knees were common around 1800, and can be seen today on the *Victory* at Portsmouth.

Extra bracing was provided at the after end of the gundeck. A large inverted knee, known as the knee at the head of the sternpost, ran along the centreline of the ship from the sternpost over three of the beams of the gundeck, and was recessed into them. In addition, knees were placed between the wing transom and the sides of the ship, a few feet above the level of the deck.

The Deck Planking

The planking of the deck was laid in straight lengths, running fore and aft; though there are signs that sometimes it was tapered and curved towards the bow and stern to allow for the narrowing of the hull, but this would have entailed extra work and was not the most common. Before the planking was put on, the 'waterways' had to be fitted. These were pieces of timber fitted round the sides of the decks. Because they followed the shape of the ship, they had to be cut to shape rather than just bent. In cross section, they formed the corner between the deck plank and the side planking of the ship. Gun deck waterways on a 74 were 5½in thick at this corner, but bevelled down to 4in on the adjacent edge so that they would meet the deck plank on the same level. The waterways were intended to keep the water on the deck and away from the frame timbers.

The planking itself was 4in thick and was laid on the four step butt system and nailed to the deck. Recesses were provided in the plan where the nails were to be used, and after the nail was hammered in the recess was covered by a small piece of timber, so that the deck presented a smooth wooden surface.

Two nails were used at each deck beam, and a trenail in each ledge.

The planks of the orlop deck were rather different. In midships, where the cables were stowed when at sea, the planks were cut in short lengths, and recesses were cut in the upper edges of the deck beams. The planks were laid between them, but not nailed. Thus they could be removed when required, to allow the water from newly raised cables to drain into the hold.

An additional system of support for the decks was provided by pillars, mostly placed under the centre of each beam. They usually had square lower and upper parts, but their central parts were turned on a lathe which resulted in one of the few decorative features in the inside of the ship. The pillars in the hold were placed on the keelson or the floor riders, and supported the beams of the orlop. They were 9¾in square at the lower end, and 1in less at the upper. Those sup-

A ship under construction at Deptford in the 1770s. The floor timbers have been crossed, the stern frame raised and the first two frames raised. Other frames have been assembled beside the slip. (The Science Museum)

porting the gundeck were of the same size, while those supporting the upper deck, quarterdeck and forecastle were slightly smaller. At gundeck level and above, many of the pillars had to be placed off the centreline to avoid being in the way of the hatches and partners.

Bitts and Catheads

The 'bitt pins' were fitted forward in the hull. Despite the suggestion of flimsiness in their name, they were substantial pieces of timber, placed vertically through the lower deck and anchored securely in the timbers of the hold. Four were fitted, in two pairs, one just aft of the foremast and the other some way behind that. Each was braced by an inverted knee or standard, in the angle between the pin and the gundeck. A strong vertical timber known as the cross piece was fitted between each pair of pins and the whole system provided strong points where the anchor cables could be secured when in use. Smaller bitts were placed around the masts for securing the running rigging

The 'cat heads' were also part of the anchor gear. Two were fitted, one on either side of the forecastle, and each was a strong piece of timber, 17½in by 16½in in cross section, with one arm projecting a few feet from the forward corner of the forecastle, and the other against the inside of the forward bulkhead of the forecastle, bolted securely to the foremost beam of the forecastle deck. It was cut with holes for sheaves at its outer end and was supported by a knee which was fitted against the side of the ship. It served as a kind of crane to lift the anchor clear of the side of the ship when it was raised out of the water.

The Order of Construction

It is easy to imagine the ship being put together systematically, with one stage being completed before the next was started, and surviving models, paintings and printed books tend to imply this. The keel would be laid, the stempost and sternpost erected and the frames put up; the planking of the outside would be put on and that of the inside up to the orlop clamp; the orlop deck beams would be laid, and planking of the side would proceed deck by deck, with the beams being laid for each deck in turn. In practice the process was much less regular. Work at the stern might progress faster than in midships, or it might stop altogether as gangs of shipwrights were taken away for more urgent tasks. There might be difficulty in finding timbers of a particular shape, or work might be interrupted by rain. Details of the actual construction of real ships is extremely rare, but the *Progress of Works* book kept at Portsmouth Dockyard in the 1740s[1] gives considerable insight into the rebuilding of the *Sunderland* from 1742–44, during the War of Jenkins' Ear, (at this time rebuilding was a purely administrative term, and in practice meant the same as building a new ship). Stages of construction overlapped with one another in quite surprising ways. The sternpost was indeed raised with some of the transoms fitted, but others were left out and only fitted much later. The planking began with the bottom, before the wales were started; the strakes under the wales were put on after the main wales, and the two parts of the planking met in the middle, halfway between the wales and the garboard strakes. The planking of the inside began at the lowest point, at the strake next to the timbers; but this stage was still being completed at one end of the ship, while the orlop deck and even the gundeck beams were being laid at the other. Most of the timbers of the frame seem to have been raised in one piece as would be expected, but some of the toptimbers were still being fitted even as the sides were being planked up as far as the gunwales in the waist.

A typical week's work, ending on 10 January 1742/3, shows how the different stages of construction might overlap;

> In hold – fayed and fastened two transom knees.
>
> Orlop – choked, fayed and fastened two lodging knees and 24 standards, and fayed the filling transom between the beams.
>
> Gundeck – let down 20 carlines
>
> Quarterdeck – brought on and fastened 5 strakes of clamp on each side
>
> Forecastle – brought on and fastened the upper strake of clamp on each side
>
> Without board – trimmed and put in their places 35 short timbers over the upper deck ports in the quarters. Brought on, set and fastened the strake of sheer rail and shut in between it and the waist rail on each side. Trimmed 11 quarterdeck beams and 6 forecastle beams.

One hundred and eighty-six shipwrights and twenty joiners were employed on the ship that week.

Clearly procedures must have changed in the 1770s, when stricter rules were applied about allowing the hull to stand in frame; at the very least, the frame must have been completed before any planking was put on. But the building of the *Valiant* must have been closer to the earlier practice as exemplified by the *Sunderland* for she too was built quickly in wartime, and the order of her building must have been haphazard.

11

Preparing for Launch

As the hull of the ship neared completion the work of the shipwrights tended to decline and there was more work for the other trades. Joiners would fit the cabins and bulkheads, and perhaps the decorative carvings of the ship. Caulkers would make the structure watertight, and painters would apply protective and decorative coatings to it. There had been some attempt to cut down on the decoration of ships during the previous sixty years and the days when each of the upper deck gunports had been surrounded by a carved wreath were past; but ships could still be described as highly decorated, with a large figurehead and supporting rails and with extensive carvings all round the stern. In1795 there were further cuts and carvings were reduced to a more austere level; but in 1759 ships were still well decorated and to a style which, though not laid down in any order, was fairly standard throughout the navy.

The Head

The most prominent decoration of any ship was her figurehead. In the early part of the eighteenth century all British warships, except the largest First Rates, had a carved

The Royal George *ready for launch on no.1 slip at Chatham in 1788. The corner of the Commissioner's House can be seen on the extreme left and this is the only building still remaining. The other buildings in the background, part of the ropemaking complex, were replaced by others by 1792 and these are still in use today.*

The Valiant *in 1790, showing the figurehead and bow decoration. (National Maritime Museum)*

The quarter gallery of a 74. (From Falconer's Marine Dictionary*)*

lion. This began to change around 1745, and by the 1750s most ships had individual heads which reflected, to a greater or lesser extent, the name of the ship. There is some doubt about the figurehead of the *Valiant*. The paintings of the ship at the Battle of Havana undubitably show her with a lion figurehead, and this is not unlikely as the lion could serve as a symbol of courage. On the other hand, Serres, who painted the pictures, shows lions on other ships as well and so perhaps this was just his habit. A print of 1790 shows the ship with a mythological figure, presumably Hercules, but this is not conclusive either, for the ship had been extensively repaired in the previous thirty years and it would not have been surprising if her figurehead had been changed.

The figurehead was placed on the end of the knee of the head, with its legs running down the sides. It was backed up by a light structure of rails and supports which were curved and reflected the shape of the head. All the rails had a deep downwards curve towards their centres though they were straight when seen in plan view. The uppermost rail curved down from the top of the head, and then up again to become almost vertical where it joined the main hull just forward of the cathead. The second rail was shorter than the others and ended on a level with the knight heads of the hull, just past its lowest point. The third curved sharply upwards after meeting the hull and merged with the supporter of the cathead. The lowest rail started at the after side of the figurehead, curved down to meet the hull and ran almost horizontally to form the upper edge of the hawse with the hawse holes under. A shorter rail began at the lowest part

The stern of the Valiant *in 1790.*
(National Maritime Museum)

of the figurehead, and ran parallel to the previous one to form the lower edge of the hawse. The area between these two rails, forward of the stem-post and against the sides of the knee of the head, was heavily decorated with bas-relief carvings representing martial or regal themes and was known as the 'trail board'. Aft of that were the hawse holes which were much more functional and under the holes themselves were surrounded by a timber known as the 'navel hood', and on top of that was a thick piece of timber known as the 'bolster', intended to protect the hull from wear and tear caused by the cable.

The supports of the rails were vertical, or almost vertical, in profile and V-shaped when seen from ahead, with the foremost ones having a much sharper angle than those aft. All the rails and supports were carved with mouldings. In addition to decorating the fore part of the ship, the head served a much less glamourous role: it provided the main lavatory accommodation for the crew and so was fitted with a structure of gratings to provide an elementary deck; there were 'seats of ease' for the men to sit on, and trunking to lead the waste down through the rails.

At the bow the structure of the hull ended quite abruptly above the level of the upper deck, with a partition known as the 'beakhead bulkhead'. This was simply built of light timber over light frames and at Trafalgar, in 1805, it was found to be one of the weak points of a ship's construction – ships sailing into battle at right angles to the enemy's fire could suffer damage because of the lightness of the structure, and soon afterwards it became common to continue the structure of futtocks and planks all the way up to the quarterdeck; but in 1759 the beakhead bulkhead was unquestioned. It was decorated with mouldings in the same style as the beakhead, and pierced with doors to allow the seamen access to the head. It had amenities of its own, for on the outside corners of the bulkhead, partly projecting over the water, were built two 'round houses' – small, circular, enclosed lavatories used by certain classes of warrant officer.

Quarter Galleries

Decorative projections known as quarter galleries were fitted on each side of the stern with their after sides in line with the flat of the stern. Like the bulkhead, the quarter galleries served a sanitary function for the captain and the wardroom officers. Seats with round holes were provided in the galleries with lead pipes leading the waste out through the lowest part of the gallery. Since the paths of such pipes would have been curved, it must have been very difficult to keep them clean.

The actual construction of a quarter gallery was very light, and contracts have little to say about it. It had the same number of rows of windows as the ship had gundecks, hence two on the *Valiant*. On a two-decker the lower row was on the upper gun deck and the higher one on the quarter deck. The most important single timber was the quarter piece, which projected from the quarter of the ship, almost in line with the flat of the stern, and formed the after surface of the gallery. The sides of the gallery were rounded in shape, and divided by five horizontal rails; one above and one below each tier of windows, and another at the level of the counter of the stern. Vertical spacing was provided by short straight timbers, which also served as the mullions of the windows where appropriate. The gallery was at its widest at the rail under the lowest tier of windows. Above that it was narrowed in stages, by means of straight lines at the windows, and curves in the spaces between them. Under the rail of the lower windows, the gallery reduced again to the lowest of its rails, and then tapered very sharply to come almost to a point, just above the level of the gun deck.

Stern Decoration

The stern of the ship can be divided into several areas, each with its own form of structure and decoration. At the lowest level was the lower counter, just above the wing transom, with its concave curve, two gunports and the helm port transom, but no decoration

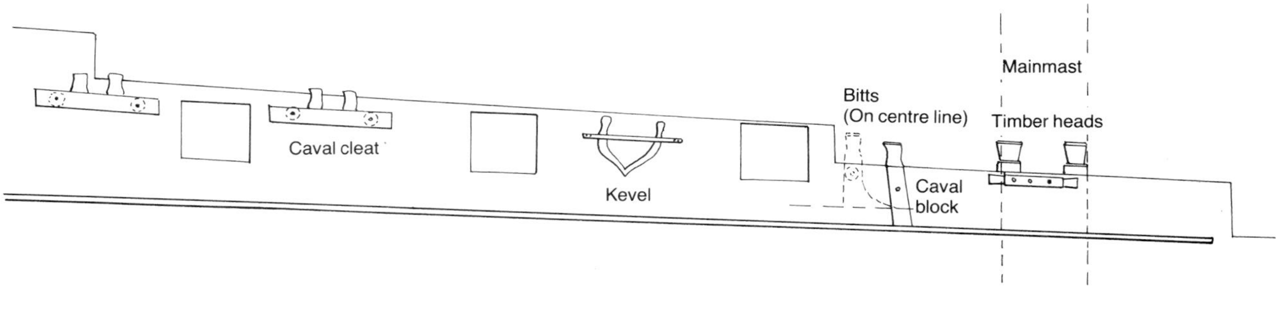

Belaying points on the quarterdeck of the Valiant.

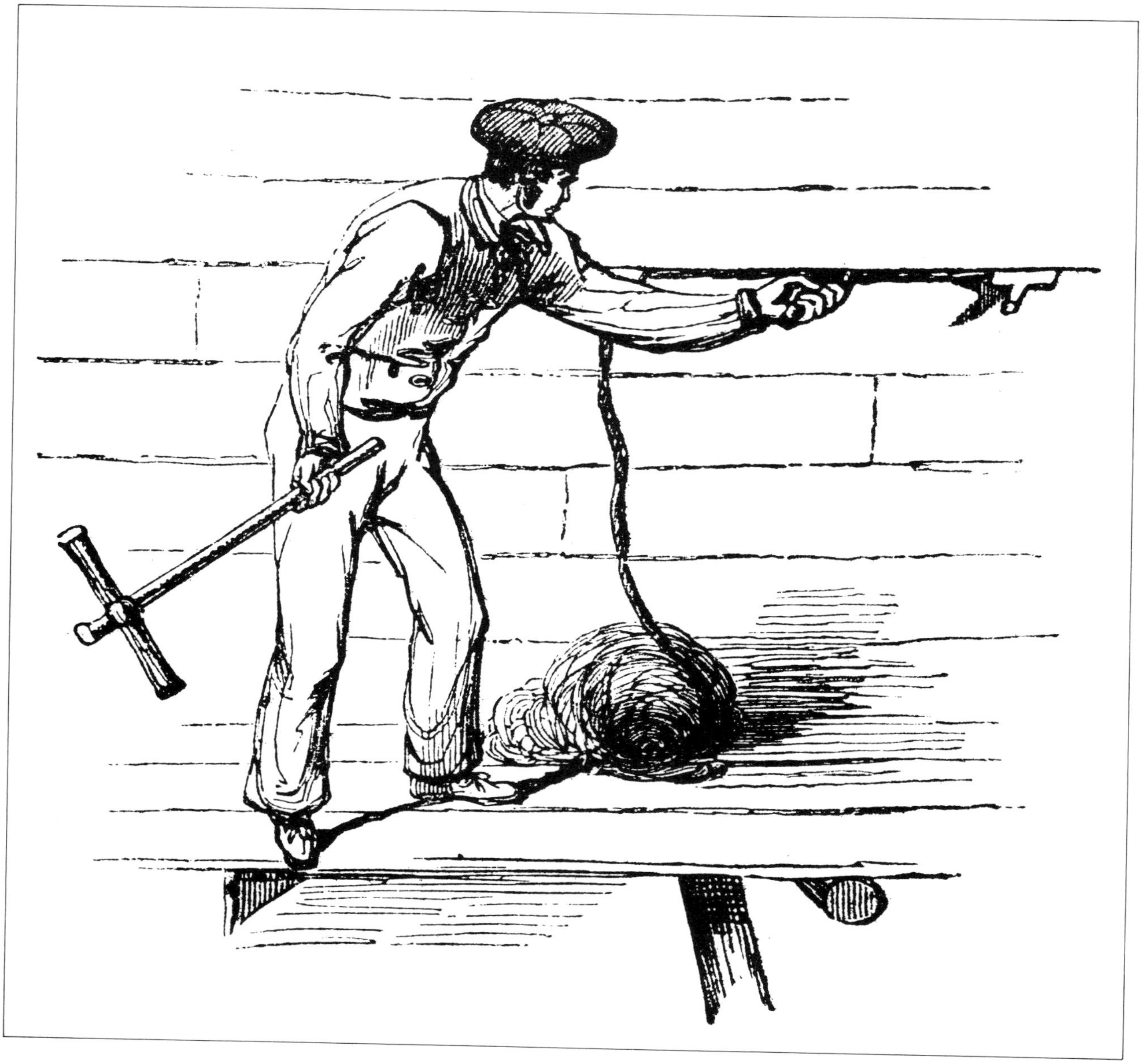

A caulker at work.
(From Days at the Factories, *1843)*

except what might be painted onto it. Above that came the upper counter, separated from the lower one by a moulded rail, and otherwise free of decoration. The counters were made simply, by light planking over the lower parts of the counter timbers. Above the rail which formed the upper edge of the upper counter were the 'stern lights' – the windows of the wardroom. The basic position of these was determined by the counter timbers, and these formed the cores of the mullions of the windows. The stern lights were on a level with the rear windows of the quarter galleries, and formed a line with them.

The counter timber, except the side counter timbers, ended at the level of the quarterdeck, just above the windows. The deck itself extended a little way beyond the windows, to form the basis of the stern gallery. The outer edge of this gallery was curved, and the deck was finished off by a moulded rail. A balustrade was placed around the gallery, and it continued across the after face of the quarter galleries. The top part of the stern was finished off by a stout piece of timber known as the 'taffrail'. It formed an arch over the stern gallery, joining the quarter pieces on each side. Just under the taffrail was a concave 'cove', bringing it down to the level of the poop deck. Between the poop deck and the quarterdeck was the screen bulkhead – a light, removable partition, with doors and windows which formed the rear face of the captain's cabin and gave him access to the stern gallery.

The stern, including the quarter galleries, attracted more decoration than the bow, though there was no single feature as dominant as the figurehead. The mullions of all the windows had carved wooden pilasters, while the balustrade of the stern gallery was supported by pillars, and the pattern of these was repeated across the rear and sides of the quarter galleries, in the form of artificial half pillars. The cove under the taffrail was carved in bas-relief and often formed a counterpoint to the taffrail itself. The lower part of the quarter gallery ended with scrolls or some other purely decorative motif.

These were relatively minor decorations. Much more dramatic were the carvings which adorned the grand arch formed by the taffrail and quarter pieces. A fully carved figure was placed on the corner of each quarter gallery with his feet resting on the end of the lower rail of the stern gallery and his head halfway up the windows of the gallery. Leaves were used to decorate the arch of the quarter gallery, and another motif, possibly a bird, covered the join between the quarter gallery and the beginning of the taffrail. A coat of arms was carved in the centre of the taffrail, supported by reclining figures, possibly an angel or a mermaid, on each side.

Belaying Points

While the decorations were being carved and put in place, the shipwrights were fitting much more functional parts to the hull and decks. Various points had to be provided to make fast, or 'belay' the ends of the ropes of the running rigging. Some, such as the bitts round the masts, were an integral part of the construction of the ship and would have been fitted earlier. Even more basic were the 'timber heads' round the forecastle deck; these were the ends of the timbers of the frames which were shaped and used as belaying points. Others were simply nailed or bolted in place, mostly along the sides of the ship. The most common method of

The painted frieze on a model of the Bellona.
(Author's photograph, courtesy of the National Maritime Museum)

A ship being floated out at Deptford. On the dockside to the extreme left, a large team of men hauls on ropes, while spectator craft fill the river. (National Maritime Museum)

A model of the Victory *showing the ship ready for launch, with the launching cradle fitted and the props supporting the hull. Though the details are accurate for such a launching, the* Victory *was, in fact, built in a dry dock (probably the same one as the* Valiant*) and floated out.*
(National Maritime Museum)

securing ropes was to the 'belaying pin rack' – a shelf fitted to the side of the ship, drilled with holes for wooden pins, each of which would hold the end of a rope, or more than one if necessary. This kind of rack was certainly in use in 1759, and one was recovered from the wreck of the *Invincible*; but it was less common than it was to become in the nineteenth century. Some minor ropes were belayed to small cleats on the shrouds while others were made fast to the timber heads which formed the ends of the hull timbers at the forecastle and quarter-deck. More important ropes, such as sheets and halyards, were belayed to stouter pieces of timber, bolted securely to the ship's sides. Several types were used, though they are not always easy to identify and describe; nautical dictionaries are not completely consistent or comprehensive, and possibly the vocabulary changed in the course of the eighteenth century. A 'kevel' was approximately V-shaped, with the ends of the 'V' protruding from a horizontal bar. A 'range' was simpler in form, being a horizontal piece of wood bolted through its middle to the ship's side, and with its ends shaped so that a rope could be turned round them. Ranges were used for the sheets of the lower sails. Horizontal timbers, with two protruding vertical pieces, can be seen on many ship's plans; these were possibly known as 'caval cleats', or 'staghorns'. Also to be seen are pieces set almost vertically against the sides of the ship, with a kind of neck near the head, to turn a rope around. These were perhaps known as 'caval blocks'.

The sheets of the lower sails had to pass through the sides of the ship at the level of the upper deck or quarter deck and spaces were left for this during the construction. Sheaves were fitted at this stage, to allow them to move freely.

Caulking and Paying

It was now time to caulk the ship. The process was defined as driving 'a quantity of oakum, or old ropes untwisted and drawn asunder, into the seams of the planks, or the interval where the planks are joined to each other in the ship's decks or sides, in order to prevent the entrance of water. After the oakum is driven very hard into these seams, it is covered with hot, melted pitch or resin, to keep the water from rotting it'.[1] Small gaps of about a quarter of an inch were deliberately left between the planks, in order to receive the caulking. The caulker was a tradesman in his own right, with tools such as the caulking iron and hammer, with which he drove the oakum into the gaps. The entire hull was caulked as well as the decks.

The hull below the waterline had to be treated to protect it from the effects of immersion in sea water. There were two main problems: the weeds and barnacles which might cling to the hull and reduce its speed, and the shipworm which could enter the wood and eat its way through and reduce the ship to a wreck within months, if unchecked. Salt water itself, of course, acts as a preservative. To protect against the worm, ships were 'sheathed', the underwater hull being covered with a mixture of hair and tar and then with light fir boards, a quarter of an inch thick, which were nailed on top. The *Valiant* was not sheathed in 1759 because worm was only found in the tropics and the *Valiant* was not to be sent there. However, she was sheathed in 1761, before being sent on the Havana expedition. In later years, it was found that copper was far more efficient than wood sheathing because it hindered the growth of weed but it was more expensive and the *Valiant* was not coppered until 1780.

For ordinary service in home waters, ships of the 1750s had their bottoms covered with 'stuff' to deter marine growths. In the past two compositions had been used: 'white stuff' which was a mixture of whale oil, resin and turpentine and 'black stuff', a mixture of tar and pitch. After 1737 a mixture of tar, pitch and sulphur was used, to produce 'brown stuff' and this was probably used on the *Valiant*, though white stuff was still in occasional use.[2]

Paying and Painting

Above the waterline, the hull was 'payed' rather than painted, below the level of the main rail. To pay was 'to daub or anoint the surface of any body, in order to protect it from the injuries of the water, weather, etc'.[3] This was usually done with a mixture of tar and oil which gave a partly transparent finish through which the natural wood colour could still be seen. Above the main rail the hull was painted, usually dark blue or red, for the sake of decoration. Models suggest that this area, along with the counters of the stern, was further painted with a frieze in gold; but it is not certain that this was always done to the real ship, especially under the pressures of wartime. The figurehead and other decorations may have been painted in gold leaf though yellow ochre was much cheaper and far more common.

Cabins

The joiners worked to install the internal fittings of cabins and partitions; though much of this work could be done once the ship was afloat, and probably was in the case of the *Valiant*. The 'establishment of cabins' was regulated by orders of 1757. Most of the officers' cabins, especially those in the wardroom, were to be made of canvas, and such work was probably done by the ship's carpenters and sailmakers rather than the dockyard personnel. Other parts were more substantial; the captain's cabin needed its screen bulkhead, as well as the internal partitions which divided off his sleeping and day cabins, and the clerks' and masters' offices just forward. On the upper deck, the wardroom needed a bulkhead, as did the cook room forward. The gundeck had no permanent bulkheads, but the orlop deck was very different. Aft in the 'cockpit', permanent cabins were erected for the surgeon and purser, along with store rooms and a dispensary. Forward were cabins for the carpenter and gunner, and a rather elaborate system of store rooms for the three 'standing officers', the gunner, purser and surgeon. Among these was built a passageway to the magazine in the hold, with a separate passage to the light room where the lantern which lit the magazine was tended. The hold was subdivided in other ways. The main magazine was forward, with the 'filling room' for making and storing cartridges close by. Aft of that was the main hold, undivided until near the stern, except for the 'pump well' built around the mainmast, and the shot lockers placed close to it. The after part of the hold was divided into the fish room (in order to keep the smell from contaminating other goods), the 'spiritous liquor room', and the bread room right aft. The partitions of the captain's cabin and the wardroom were light and removeable, though a good deal of care was needed in their construction; those of the orlop and the hold were generally rougher and stronger, because the storerooms and compartments needed to be secure against theft.

Launching

The launching of a ship represented the real beginning of her active life. A modern warship might spend years in a shipyard after her launch, being fitted with engines, electronics and weapons systems. A ship of the eighteenth century, once afloat, could be prepared for war in about three weeks if the labour and materials were available. True, many a ship, such as the *Victory*, might spend years laid up in ordinary in peacetime; but as long as her hull was kept in good repair she was almost ready for action, and she had deterrent value as part of the strategic balance.

As the ship approached completion, the dockyard officers chose a suitable day for the launch. High tide was necessary, and for a ship of the line it had to be a spring tide, occurring fortnightly. The Admiralty was informed of the intention and usually agreed without question. A structure was erected around the ship which replaced the props, shores and scaffolding which had supported her during construction. Two wooden rails were built on each side of the keel, sloping downwards and into the river. A launching cradle was built under the bow and stern of the ship. This was made in four parts, one under each side of the bow and stern. Each part consisted of a strong horizontal timber, with lighter pieces rising from it, each shaped to conform to the shape of the hull at that point. The same cradle could be re-used for many ships of similar size, and it was not necessary to make a new one except for very unusual ships. The cradle was designed to run on the two wooden rails, and would support the hull as it slid into the water.

After the cradle had been fitted, the rails were greased to allow it to run smoothly. Wedges were driven under the rails to raise the ship slightly, and take some of the weight off the blocks. Shortly before the launch the blocks were removed splitting them if necessary and starting with the aftermost one.

Some props were kept under the stern until soon before the launch. These were knocked away, and the ship was only supported by ropes leading from the hawse holes to bollards at the head of the slip. These were cut, and the ship slid into the water, to the cheers of the onlookers.

Of course, none of this applied to the *Valiant*, for she was built in a dock rather than on a slip and she would have been supported by horizontal props leading from the dockside to the side of the ship. The sluices would have been opened on the rising tide and as the water rose in the dock the ship would begin to float and the props could then be removed. Then teams of men would haul the ship out into the river, using ropes attached to the bow. It was a much slower process than launching and less dramatic and it is not clear what was considered the official moment of launching – when the ship first began to lift, or when she was fully afloat, or when she had left the dock.

Launching Ceremonies

Not much has been recorded about the launching ceremonies of the eighteenth century. Nothing is known about the *Valiant's* as there was no local newspaper in Kent to report the matter, and the national press took little interest. At the beginning of the century it had been customary to make a gift of silver to the master shipwright, but this seems to have died out by the middle of the century. The custom of breaking a bottle of wine over the bows is recorded in 1802,[4] and was considered old by that time. The practice of having a new ship launched by a woman was not yet established and merchant ships were often launched by their owner or the prospective captain, but this is not likely for a naval ship. Probably a dockyard official, such as the commissioner, carried out the ceremony. The Master Shipwright would certainly be there, and he would have to work hard in supervising the shipwrights. One custom is well documented. Ships to be launched had not yet had their masts fitted, but light flagstaffs were placed in the mast positions and decorated with enormous flags. A small Union Jack flew from the bow; the Admiralty flag, an anchor on a red background, from the foremast position; the royal standard from the mainmast; a larger union flag from the mizzen; and an ensign (usually the red ensign if paintings are to be believed) flew from a flagstaff at the stern. Contemporary pictures suggest that a flotilla of boats of all shapes and sizes would have watched from the river, but the best experience would undoubtedly have been for those who were allowed to be aboard the ship as she ran into the water. The launch being completed, no doubt celebrations were held afterwards, perhaps with the dignitaries being invited to the Commissioner's House, but there is no direct evidence for this.

On the 10 August the officers of Chatham Dockyard wrote to the Navy Board reporting that the launch had taken place successfully. She was moored at buoys in the river in preparation for fitting out.

12

Masts and Yards

After her launch a new ship might not be fitted for the sea for some time. If she took to the water in peacetime, she might well be laid up in ordinary for many years. This happened to the *Valiant's* successor in the drydock, the *Victory*, which was launched in 1765 but did not sail until 1778. A ship in ordinary would be fitted with cabins for her standing officers – the boatswain, carpenter, purser, cook and gunner, and a small number of servants. She would have a wind sail, or some other means of allowing air to her timbers. Otherwise she would be an empty shell, without guns, stores, masts or any of her fittings. Even so, she might be usefully employed. The 90-gun *Glory* was laid up at Chatham after the Seven Years War and her purser, William Falconer, used the time to compile his famous *Marine Dictionary*, published in 1769. 'The captain's cabin was ordered to be fitted up with a stove, and with every addition of comfort that could be procured; in order that Falconer might be able to enjoy his favourite propensity without either molestation or expense'.[1]

There was no prospect of the

A view towards Upnor Castle in 1759, showing part of the newly-built dockyard defences. The ship in the river may well be the Valiant, *as she was fitted out at exactly this time and place. (British Library)*

The masthouses at Chatham, newly restored in 1989.

Valiant spending her first years in peace. As she floated out in August 1759 the conduct of the Seven Years War had reached a critical phase. The war had begun badly for Britain with the loss of the vital Mediterranean base of Minorca, leading to the execution of Admiral Byng. Since then there had been victories in the overseas empires; Clive had won the Battle of Plassey in 1757, while Boscawen had defeated the French fleet at Lagos, and General Wolfe had taken Quebec. But the main French fleet remained undefeated in Brest harbour, and England was threatened with invasion. A new ship of the line, especially one as large and powerful as the *Valiant*, would be an important addition to the fleet. Furthermore, there are signs that the *Valiant* had already caught the eye of one of the navy's most distinguished and influential young captains. Augustus Keppel, a close friend of Lord Anson, had praised the sailing qualities of the *Invincible* ten years earlier, and at some stage the new *Valiant* was set aside as his command. Every effort was to be made to get the ship into action as soon as possible.

Types of Masts

A ship of the line invariably carried three masts. The largest, the mainmast, was placed approximately in the centre of the ship. The next largest, the foremast, was close to the bow, only 12ft from the beakhead bulkhead. The third was the mizzen mast, set about halfway between the mainmast and the taffrail. The gap between the mainmast and the foremast was much greater than that between the main and the mizzen. The mainmast was aft of the centre of the ship and helped balance her as the sails on the mizzen were much smaller than those of the foremast. This space also helped when tacking ship, as the sails of the foremast had to be braced round in the opposite direction to those of the mainmast to help carry the ship through the eye of the wind; without this space the yards would have fouled one another.

Each mast was made in two or three sections, overlapping and joined by means of a system of 'caps' and 'tops'. The mainmast proper began just above the keelson of the ship, and passed through the various decks, and protruded more than 80ft above the quarterdeck. Its uppermost section was square where it overlapped with the section above, the main topmast. The upper part of that mast was also square in section and overlapped with the lower part of the main topgallant mast which itself was finished off with a flagpole and button. The foremast was similar, consisting of the foremast proper, fore topmast and fore topgallant. The mizzen mast, in this period, had no topgallant but merely a light flagpole in its place.

The other important spar on a ship was the bowsprit which projected out from the bow at an angle of about 30 degrees and whose principle use was for setting the foresails and spritsails and for securing the stays of the foremast. The bowsprit had a jib-boom, which bore the same relationship as a topmast to a lower mast.

Mastmaking

Masts were made in the dockyard at Chatham in the newly constructed masthouses under the mould loft; even warships constructed in the merchant shipyards had their masts made in the Royal Dockyards. There was no separate trade of mastmaker: it was merely a branch of the shipwright's profession.

Supplies of timber for masts came from North America, and at this period there was no shortage of large trees. It is likely, therefore that each of the *Valiant*'s lower masts was cut in a single piece. Twenty years later, supplies of North American timber were to be seriously interrupted, and the navy had to rely on 'made' masts – those constructed from several pieces of timber, interlocked together in a complex system. Even in the 1750s, made masts were not unusual, and so it is possible that they were used for the *Valiant*.

A lower mast was round in cross section for most of its length. It was

XXI.—Interior of Mast-house. Face p. 480.

Mastmakers at work in the nineteenth century, driving iron hoops onto the mast. This method only became standard in 1800, and before that the masts were held together by rope woldings. (From Days at the Factories, *1843)*

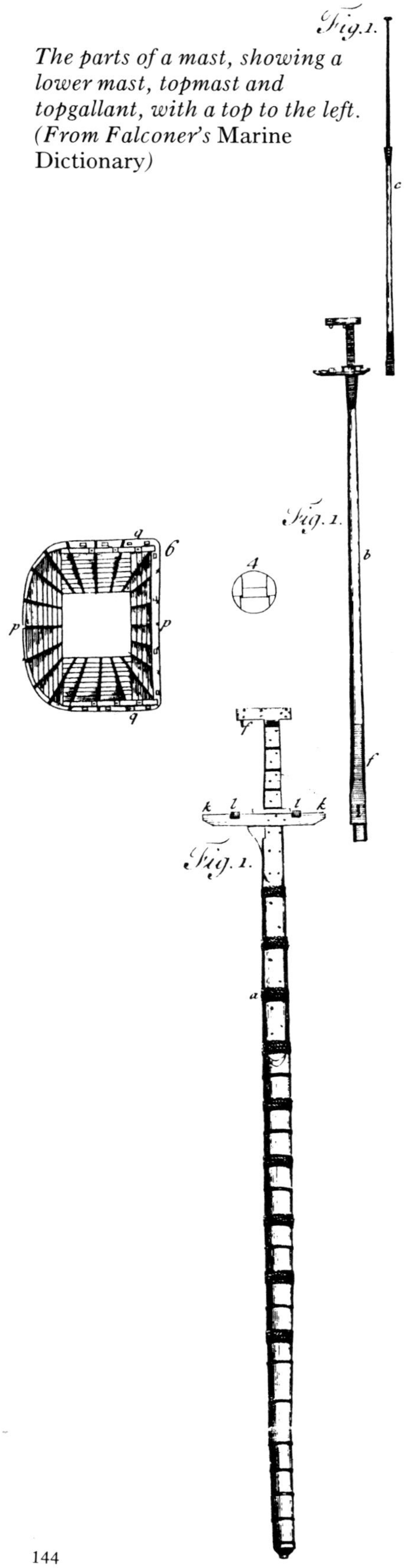

The parts of a mast, showing a lower mast, topmast and topgallant, with a top to the left. (From Falconer's Marine Dictionary*)*

A sheer hulk, lifting the lower mast of a ship into position. The figures are out of scale, and are far too large.
(From Serres' Liber Nauticus*)*

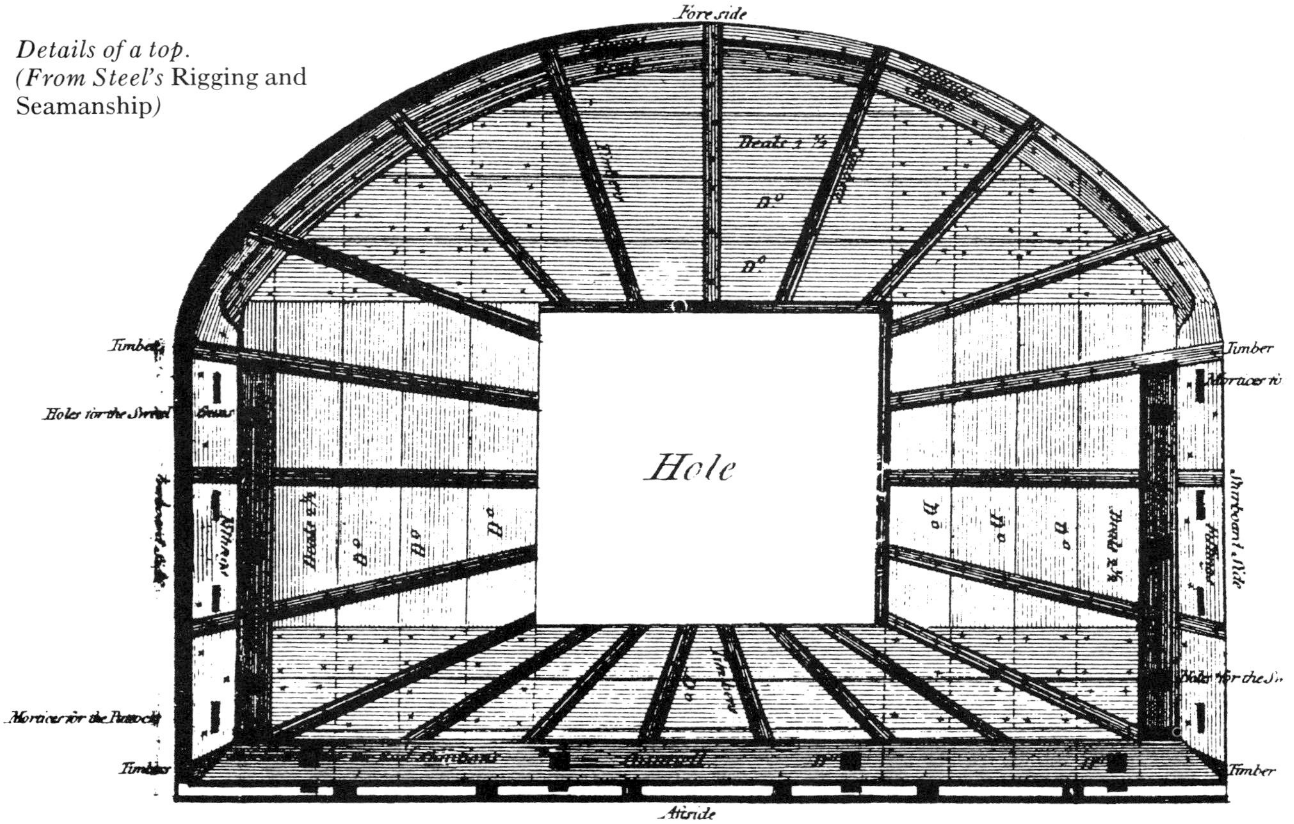

Details of a top. (From Steel's Rigging and Seamanship*)*

roughly cigar-shaped, with its greatest width at the 'partners', at the point where it passed through the upper deck. Its lowest end, the heel, had a tenon to fit into the hole in the mast step, on the keelson. The core of the mast continued its circular cross section until the 'hounds', where it began its overlap with the topmast and here it became rectangular and the side parts of this rectangle, which formed a kind of shoulder to hold the 'top' of the mast, tapered away to about halfway down the sides of the mast, forming the 'cheeks'. Above the hounds, the mast was square and tapering, and ended in another tenon at the head, for the cap. Such a mast, if constructed from several pieces of timber, probably had a central part known as the 'spindle', which itself might be made of more than one tree; and it had side pieces added to form the cheeks and sides of the mast. 'Tables and coaks' were cut in the pieces to make the joins strong. There were many different ways of constructing 'made' masts, and few universal rules. The lower mast, whether made or not, was strengthened with rope 'woldings'. These were ripes wound round the mast at intervals, and held in place by wooden bands nailed to the mast. These woldings were intended to prevent the mast from splitting. The bowsprit, which was rather similar to a lower mast in construction, also had woldings. It was circular for nearly all its length, but its outer end was fitted with a flat surface through which rigging lines would be passed, and known as the 'bees' because of its shape in former times.

Upper masts – topmasts and topgallants – were each made in a single piece and had no wolding. A tòpmast was square in section at its heel, with a small hole cut across it to insert the fid which would hold it in place. It was also cut with a fore and aft hole, for fitting a sheave which would be used for raising and lowering it. Its next

Methods of constructing yards, by tonguing and by scarphing. (From Falconer's Marine Dictionary*)*

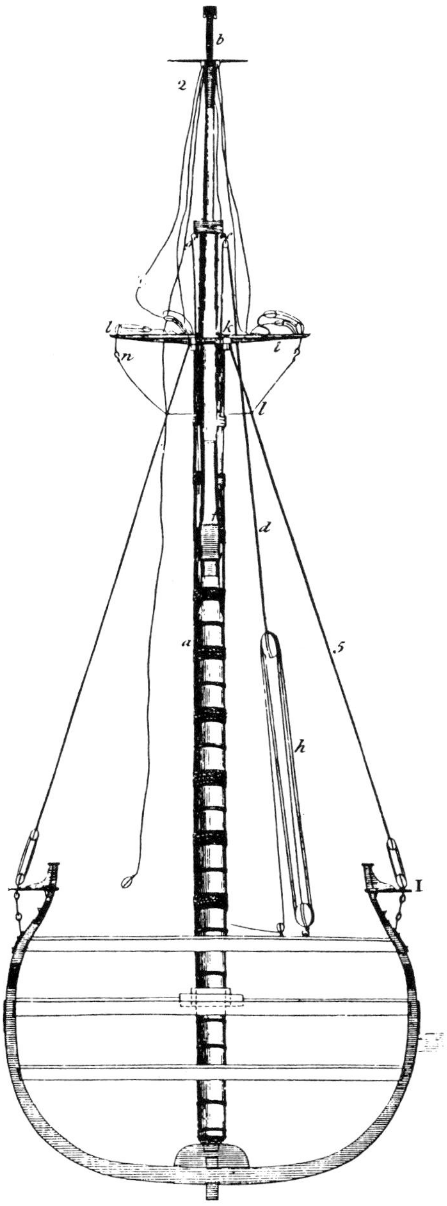

A topmast being raised into position.
(From Falconer's Marine Dictionary*)*

part was hexagonal in section, and that was the part which would overlap with the head of the lower mast. It then became round in section, and tapered until just under the hounds. The hounds themselves were also hexagonal in section, and provided a support for the 'trestle trees' of the mast. The hexagonal shape tapered under the hounds, to meet with the rounded shape of the main part of the mast. Above the hounds, the mast was square and tapering, as with the lower mast. A topgallant mast was similar to a topmast, but instead of the square section above the hounds, it had a round pole, with a button on the top which served as a flagpole.

Fitting of Masts

The lower masts of the *Valiant* were fitted by the dockyard workers, before the ship was commissioned under Captain Brett. The ship was hauled alongside a sheer hulk moored in the river off the dockyard. This vessel was an old warship with its original masts, guns and fittings removed but fitted with a set of sheers which formed a kind of crane. The mast was floated out to the ship, or conveyed on a lighter, and lifted until it was vertical. From that position it could be passed through the partners on the various decks of the ship and its heel fitted into the tenon over the keelson. The holes at the partners on the various decks were larger than the mast itself and the space between was filled by wedges; canvas was then nailed over the join. After that the standing rigging of shrouds and stays would be fitted, probably by the crew of the ship rather than the dockyard workers. The sheer hulk was also used to fit the bowsprit, which passed between the knight heads of the hull, and then through a hole in the beakhead bulkhead. A strong timber, made from two halves scarphed together, was placed at-an angle between the beams of the upper deck and the forecastle and cut with a square hole to receive the heel of the bowsprit.

The next stage in rigging the masts was to fit the 'top' and its associated structures. Straight pieces of timber called the 'trestle trees' ran fore and aft from the hounds of the mast. These interlocked at right angles with two other timbers, known as 'cross trees', which were fitted fore and aft of the mast, leaving enough space forward for the heel of the topmast. The trestle trees and cross trees together formed a kind of double cross and its central part served to hold the lower mast and the topmast together, while its outer arms supported the top itself. This was a flat structure, roughly D-shaped, with a square hole in the middle. It was constructed of planks running fore and aft and athwartships, and had a very low rail round its edges, and a series of ribs radiating from its centre. As well as helping to brace the structure of cross and trestle trees, the top provided a surface for men working on the upper rigging, and sometimes a platform for light guns or carronades. It also helped spread the shrouds of the topmast. After the cross and trestle trees were in place, the top was fitted by lifting it over the head of the lower mast, and then setting it down.

Once the top was in place, it was time to fit the topmast. On the whole this was simpler than the mainmast and no sheer hulk was needed. The topmast (which was shorter then the

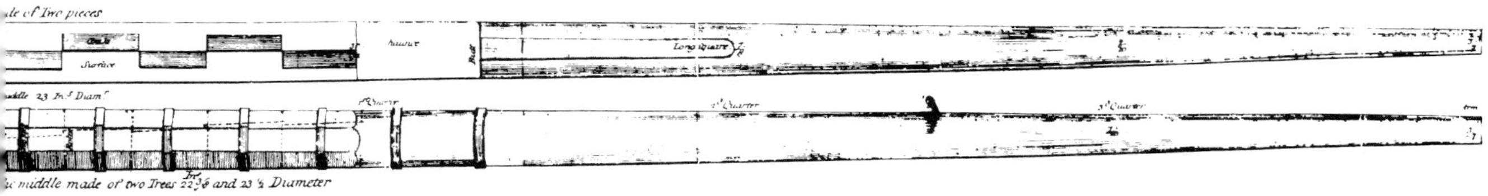

lower mast) was placed vertically under the gap in the top. The 'top rope' was made fast to a position near the head of the lower mast and then passed through the sheave in the foot of the topmast, back up to a block near the head of the lower mast, and then down to the deck where it might be led to a capstan. The topmast could then be raised by hoisting on the top rope. The upper part of the topmast passed through the 'cap', which was simply a rectangular piece of timber fitted over the head of the lower mast, and drilled with a large hole for the topmast to pass through. When the heel of the topmast was level with the top of the lower mast, it was time to insert the fid. This was simply a timber peg, put through a hole in the lowest part of the topmast to hold it in place. The raising and lowering of the topmasts and topgallants was kept relatively simple because it often had to be done at sea, on the approach of bad weather. Even when fitting out at Chatham the *Valiant* had to lower her topmasts on the 14 and 21 September when gales were expected.

At the head of the topmast was another system of cross trees and trestle trees. This was similar to the one below but no top was fitted here. The cross trees served to spread the shrouds of the topgallant mast. The fitting of the topgallant mast was similar to that of the topmast.

The jib-boom was held to the bowsprit by means of a cap placed over the end of the bowsprit. The heel of the jib-boom was laid on a block of wood known as a saddle, and was then lashed down by the heel ropes.

Once the masts had been set up, and supported by the appropriate standing rigging, it was possible to begin raising the yards and other spars. As with the construction of the ship, the order was not always as systematic as logic might suggest. The lower masts and bowsprit had already been fitted by the 22 August, when the ship was commissioned. The caps of the masts were set up by 4 September, and the topmasts were brought on board and raised two days later. The lower yards were set up next, during which there was a serious accident. 'Getting in the main yard the lashing slipped, by which the third lieutenant had his thigh broke'. Nevertheless, the main and fore yards were 'swayed up' on the 8th. The topsail yards were brought on board on the 11th, and rigging of them began. They were raised up to the tops and then set up the next day. The topgallant masts and yards were set up on the 15th and the sails came on board on the 19th. During all this time work continued on other aspects of the ship's fitting – ballast was loaded, stores were brought on board, and so on.

Yards

The yards were made in the masthouses of the dockyards. Each of the sections of mast – foremast, fore topmast, fore topgallant, for instance – had a yard associated with it, though there were some exceptions. There was no square sail on the lower mizzen but the sail above, the mizzen topsail, needed a square yard, the mizzen topsail yard, to spread its foot. The mizzen yard served a different purpose for it was hung fore and aft, and fitted with a kind of gaff sail. In effect, the lower mizzen had two yards.

The main yard of a ship was almost as long as the mast to which it was attached, but it was both lighter and simpler in construction. Its central part was usually octagonal, while the outer lengths were round and the whole tapered towards the ends. Most yards were made from a single piece of timber, but some of the largest ones had two pieces scarphed in the middle, or consisted of a single central piece with extensions 'tongued' on the outer ends.

Some lighter spars were needed for supplementary purposes. 'Boomkins' extended diagonally downwards and outwards from the knight heads in the bow, and were used to lead the tacks of the fore courses. Studding sail booms were fitted above the outer edges of the main and fore yards and could be run out in light winds to carry studding sails. The lower studding sail booms were hinged against the sides of the ship, and extended the lower corners of the studding sails in very light winds. The studding sails also had yards, attached to the top of the sails themselves.

13

Sails and Rigging

A 74-gun ship of this period needed around 20 miles of rigging, excluding the ropes used for anchor cables and their tackle, for rigging boats, and for miscellaneous purposes such as mooring lines. The thickest rigging line was the main stay, 18in in circumference and therefore about 6in in diameter. Most ropes were between 2in and 6in in circumference though some were as little as 1in. Most of this rope was made in the Royal Dockyards, in the ropeworks. The one at Chatham was built in its present form around 1792, on the site of an earlier one, and is 1,135yds long; long enough for the great anchor cable to be made up in single lengths. A ropery has occupied the same site since the early seventeenth century.

Ropemaking

Rope was made from hemp, mostly imported from the Baltic. It was cleaned and combed out by means of a 'hatchel', a piece of wood with forty iron teeth, each 1ft long, set in rows. It was next spun into yarn 1,020ft long, tarred and set aside for a few days. The yarns were taken into the main ropeworks and formed into strands. If normal rope

Ropemakers at work in Chatham Historic Dockyard, laying the rope and using the top. Two three-strand ropes are being laid. On the nearest one the ropemaker, riding the top frame, has wrapped some rope round the rope being laid in order to protect his hands as he uses the rope as a brake.

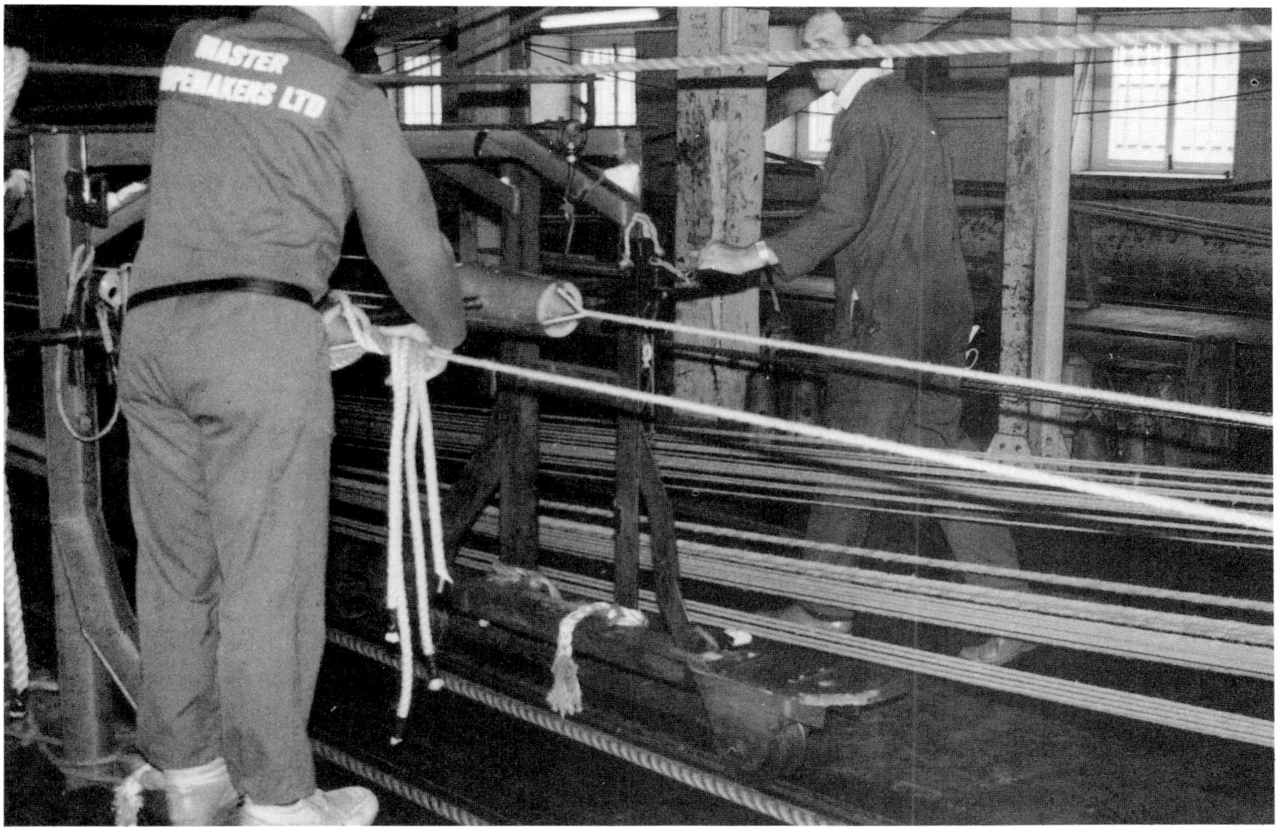

Side view of the rigging of a mainmast. No 5 represents the lower shroud, with the backstays behind. (From Falconer's Marine Dictionary*)*

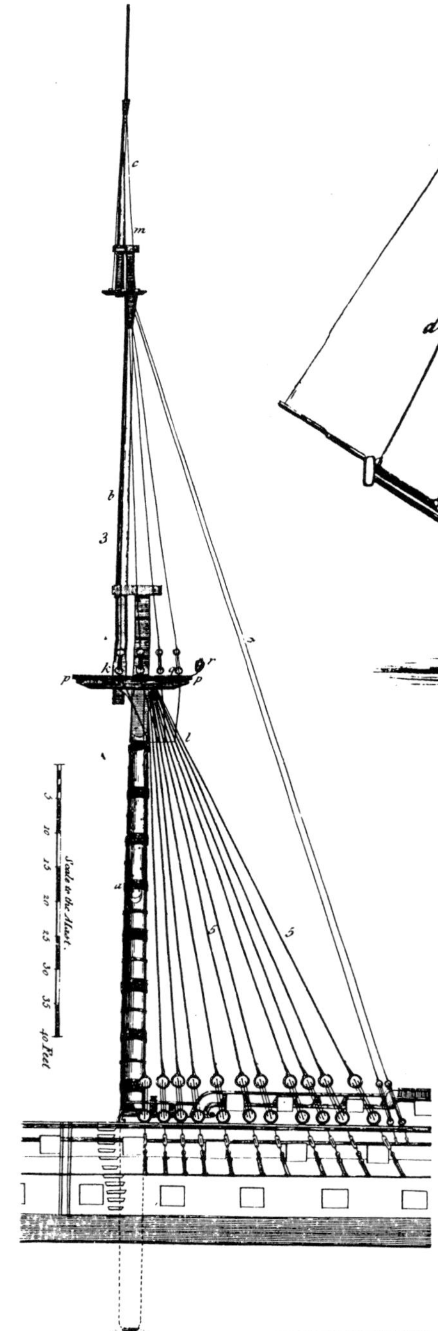

The stays of a 74-gun ship. (From Falconer's Marine Dictionary*)*

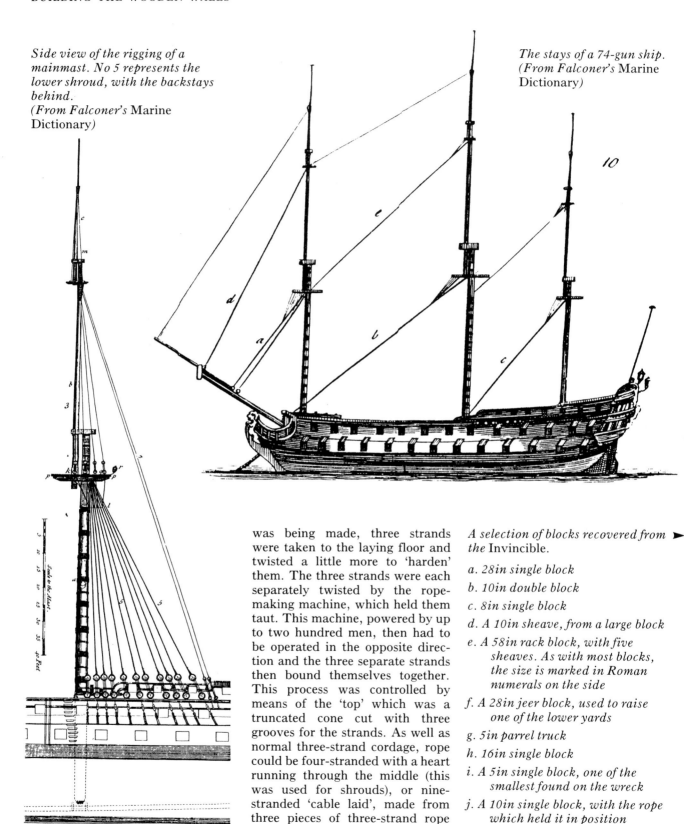

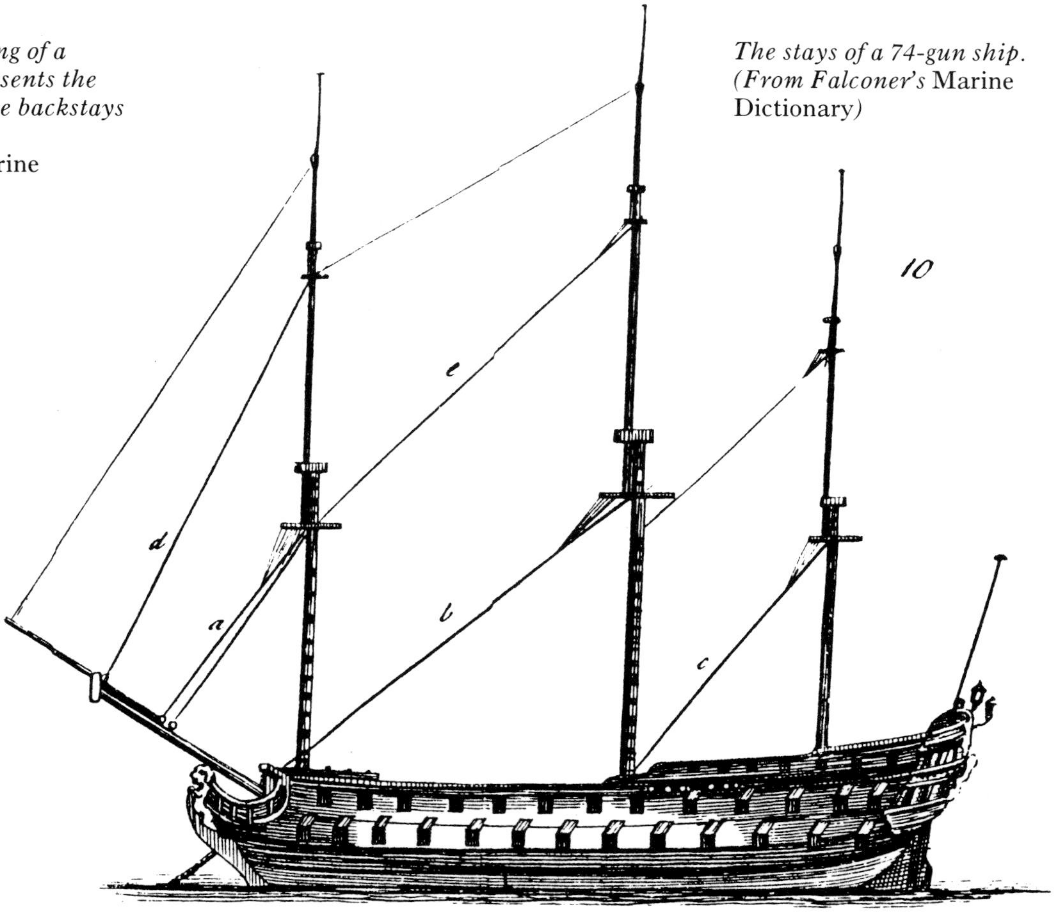

was being made, three strands were taken to the laying floor and twisted a little more to 'harden' them. The three strands were each separately twisted by the rope-making machine, which held them taut. This machine, powered by up to two hundred men, then had to be operated in the opposite direction and the three separate strands then bound themselves together. This process was controlled by means of the 'top' which was a truncated cone cut with three grooves for the strands. As well as normal three-strand cordage, rope could be four-stranded with a heart running through the middle (this was used for shrouds), or nine-stranded 'cable laid', made from three pieces of three-strand rope wound together.

A selection of blocks recovered from the Invincible. ►

a. 28in single block

b. 10in double block

c. 8in single block

d. A 10in sheave, from a large block

e. A 58in rack block, with five sheaves. As with most blocks, the size is marked in Roman numerals on the side

f. A 28in jeer block, used to raise one of the lower yards

g. 5in parrel truck

h. 16in single block

i. A 5in single block, one of the smallest found on the wreck

j. A 10in single block, with the rope which held it in position

a
b
c
d
e
f
g
h
i
j

Part of a parrel from the Invincible.

Standing Rigging

A ship was normally rigged by her crew, but the *Valiant* was severely undermanned and so the rigging was largely done by dockyard personnel and by men lent from other ships. Rigging falls into two classes: standing rigging which supports the masts and is rarely handled except when occasional tightening is required; and the running rigging which supports and controls the yards and sails and is handled constantly in order to set or take in sails and to trim them to suit the wind. The running rigging was put on without any preparation, but the standing rigging was usually tarred, and treated in other ways to protect it from the weather. Stays, for example, were 'wormed, parcelled and served'; a light rope was wound into the groove between the strands of the main rope, to produce a smoother surface, and then paper was wrapped round the whole assembly; a yet lighter line was then wound round that.

There were three main types of standing rigging. The shrouds ran from the head of each section of mast, outwards and backwards to the channels in the case of the lower masts, or to the tops and cross trees in the case of the topmasts and topgallants. Under the tops and cross trees were the 'futtock shrouds' which carried the ends of the upper shrouds inwards to meet the lower ones. 'Ratlines' were fitted to all the shrouds, including the futtock shrouds. These were ropes tied horizontally to the shrouds and provided a kind of ladder by which the seamen climbed up into the rigging. The heads of the shrouds were looped over the head of the mast in pairs. At the channels, each shroud ended round a block known as a deadeye. Another deadeye was fixed to the channels and a lighter rope, known as a lanyard, passed between them, being laced through three holes in each deadeye. The lanyards were hardened up from time to time in order to tighten the shrouds. The mainmast of the Valiant had ten shrouds on each side, the foremast had eleven and the mizzen had six. The main topmast of a 74 usually had six and the topgallant had three.

Backstays supported the masts against pressures from behind and they differed from shrouds principally in that they extended over two mast sections or more, running from the channels to the head of the topmast or topgallant. Some backstays could be loosened in order to make room for yards and sails to be trimmed and these were known as running backstays; most were set up permanently.

Forestays ran forward from the mastheads to meet the bowsprit in the case of the foremast, the deck in the case of the lower main and mizzen masts, and the masts in the case of the main and mizzen topmasts and the main topgallant. Like shrouds and backstays, they were tightened by means of a system of lanyards and passed through open blocks known as hearts.

The head of a stay was formed into a loop which was passed over the head of the appropriate mast. On topmasts and topgallants this loop was formed by means of an eye-splice. On the lower mast, the loop was formed by means of a 'mouse' and 'collar'. The collar was a small loop formed in the end of the stay, just big enough to let the other end of the stay pass through. The mouse was a kind of bulge formed in the rope, to prevent it from passing too far and tightening on the masthead.

Blocks

A ship like the *Valiant* needed about 1,400 blocks for her rigging and guns. They could be used to alter the lead of a rope round an obstacle, or they could serve as part of a block and tackle arrangement for lifting heavy weights. A block consisted of three main parts: the sheave, made of lignum vitae, the shell which enclosed it, made of elm, and the iron pin joining the two and which allowed the sheave to turn. The simplest block had only a single sheave, while others had several which were fitted side by side. A fiddle block, for example, had two sheaves, one above the other, giving the shell its distinctive shape; a rack block was used on each side of the bowsprit, and consisted of five or six sheaves in line.

Chatham Yard had only two blockmakers in 1758, mainly employed on repair, and so most blocks were made under contract. In 1759 Taylors of Southampton were experimenting with mass production and this was to culminate in Brunel's system of machine tools which were installed at Portsmouth in the 1800s. A spare set of machines was made for Chatham, but was never used as Portsmouth satisfied all the demand.[1]

Running Rigging of the Yards

Rope for the running rigging falls naturally into two different types: that used to hoist and control the yards, and that used to trim the sails themselves. The first task was to raise the yards which supported the principal sails. In the case of the main and fore lower yards, four large double blocks known as 'jeer blocks' were employed: one pair attached to the top and the other to the yard. These yards stayed aloft more or less permanently, for they were not taken down in bad weather, nor moved when furling the sail. The topsail and topgallant yards were hauled up by means of a system involving three blocks: one on each side of the trestle trees and the other in the centre of the yard.

The yard was held against the mast by means of 'parrels'. Each consisted of a set of 'B'-shaped ribs with beads, known as 'trucks', assembled in between them. The upper yards used a truss which was simply a rope which held the yard to the mast.

To keep the yard level 'lifts' were used. These ropes passed through blocks at the cap of the mast, and at the yardarms, and then were led down to the deck to be secured. 'Braces' were used to turn the yards so that the sail could be turned to the direction of the wind. Each came in two parts: the 'pendant' and the 'fall'. The pendant was attached to the yardarm, with a block at its other end and the fall passed through it, one end being made fast to the standing rigging, and the other passing through another block and leading to the deck below. The braces of the fore course, topsail and topgallant led aft to the stays of the mainmast; those of the main course, topsail and topgallant also led aft to points on the mizzen mast and topmast. The braces of the mizzen, on the other hand, had to lead forward, to the main shrouds and to the main topmast trestle trees.

A sailmaker at work sitting on his bench. To the right, a tackle is used to stretch the bolt-rope of a sail. (From Steel's Rigging and Seamanship*)*

The sails of a ship of this period. The Valiant*'s sails would have been similar except that the mizzen topgallant (no 21) and spanker (no 19) would not have been fitted in her early years.*

The sails are:

1. Flying jib

2. Jib

3. Fore topmast staysail

4. Fore staysail

5. Fore sail, or fore course

6. Fore topsail

7. Fore topgallant

8. Main staysail

9. Main topmast staysail

10. Middle staysail

11. Main topgallant staysail

12. Main sail, or main course

13. Main topsail

14. Main topgallant

15. Mizzen staysail

16. Mizzen topmast staysail

17. Mizzen topgallant staysail

18. Mizzen sail

19. Spanker

20. Mizzen topsail

21. Mizzen topgallant

T. Swaine sculp.

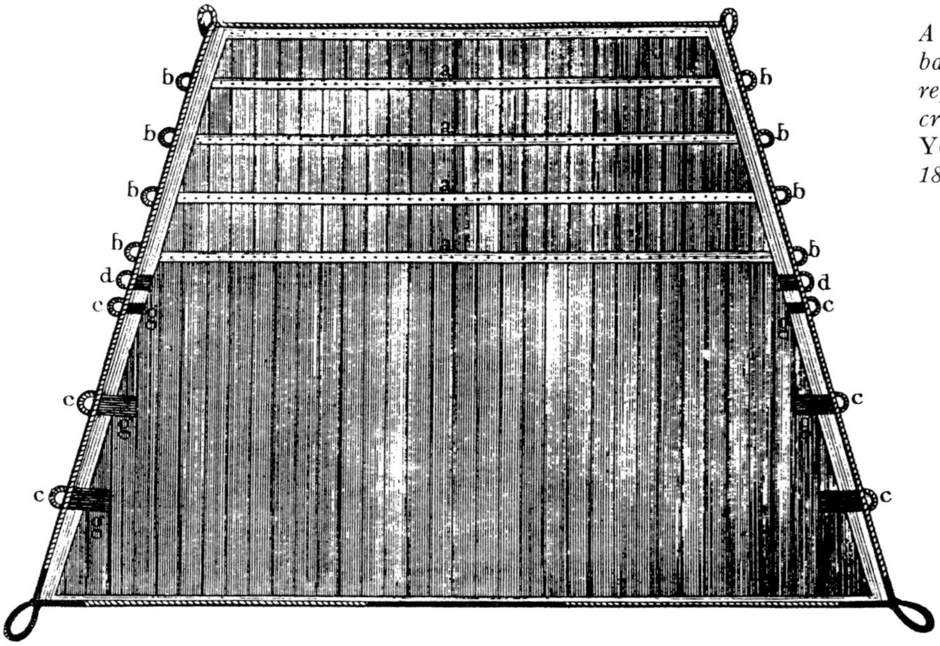

A typical topsail. (a) shows the reef bands and (b) the reef cringles; (c) represents the bowline cringles.(From D'Arcy Lever's Young Sea Officier's Sheet Anchor, *1805)*

Sailmaking

The most important sails on a ship of the line were the 'square' sails, so called because they hung at right angles to the direction of the ship. These sails provided the main motive force of the ship and were all hung from yards. The lower sails on the fore and mizzen, known as 'courses', were parallel sided, with a pronounced curve, or 'gore' on their lower edge. Topsails and topgallants were narrower at the top than the bottom, and had no gore. Square sails were also carried on the yards under the bowsprit. The spritsail itself was completely rectangular, with no gore. The spritsail topsail was rather like an ordinary topsail in shape, with its top edge smaller than its lower. Studding sails were hung from booms at the ends of the yards in very light winds and used to extend the sail area.

'Fore and aft' sails were so called because they ran fore and aft. The mizzen course was the only one supported by a yard, and was hung to the after part of it, behind the mast, rather like a modern gaff sail. There were many staysails, hung from the stays which supported the masts and these were quadrilateral in shape, with an angled forward edge to accommodate the angle of the stay itself. Jib sails were triangular in shape and hung from the stays of the foremast, occupying the space between it and the bowsprit.

Sails were made in the dockyards by skilled sailmakers. The sail loft at Chatham, still in existence and still used for sailmaking, was built in the 1720s. Forty-five sailmakers were employed there in 1758. Sails were made from canvas, in different thicknesses according to their shape, size and use. Canvas was graded by number; the thickest, used for courses and for topsails to be hoisted in heavy weather, was number 1. The lightest, used for studding sails set only in the lightest winds, was number 7. Sailcloth came in rolls known as bolts, 2ft wide and 38yds long. A bolt of no 1 canvas weighed 44lbs, and one of no 7 weighed 24lbs.

The first job of the sailmaker was to cut the canvas to appropriate lengths and shapes of the sail in question, and then sew the edges of the pieces together, using a 'double flat seam', with 108 to 116 stitches to the yard. All sewing was done by hand. The edges of the sail were turned over and sewed down to form a kind of hem, known as 'tabling'. Around all the edges of the sails was sewn a rope, the 'bolt-rope' and at the corners of the sail loops known as cringles were formed in it. Some sails, especially topsails, had additional cringles in their sides for fixing reefing tackles, and they also had 'buntline cringles' in their lower edges to attach the ropes for furling the sail. Most sails also had a row of small eyelets just under the top edge which were used for the 'robbands' (rope-bands) which would attach the rope to the yard or stay. Topsails and courses had lines of 'reef points' running acrosss them. These were light ropes passing through the sail and used to furl the top part of the sail in heavy weather. Many sails also had reinforcements in the parts most subject to wear and tear, in the form of double thicknesses of canvas.

Running Rigging to the Sails

The upper corners of a square sail were stretched along the yard by ropes known as 'ear-rings'; the lower

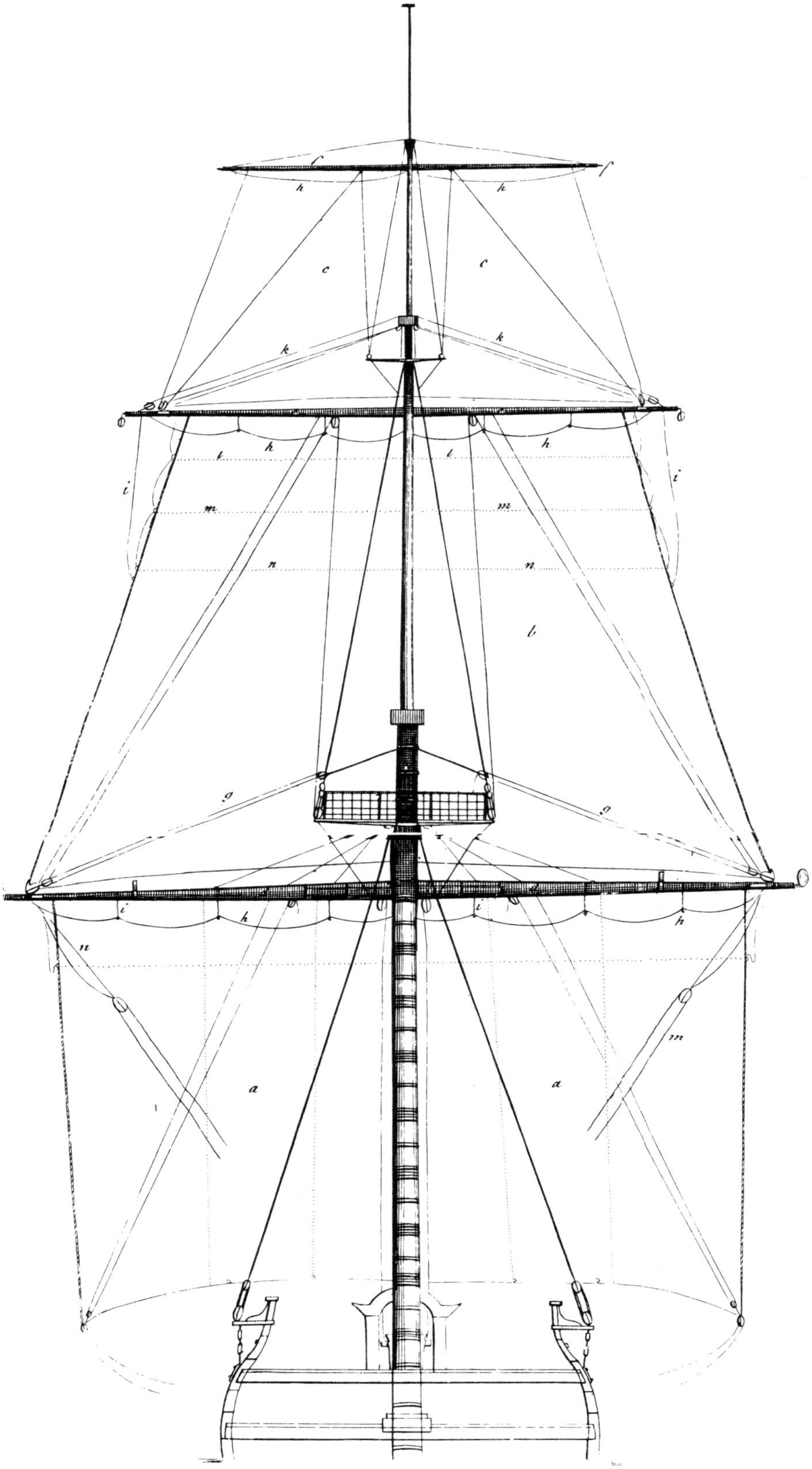

The sails and rigging of a mainmast.

m. The main braces

h. The foot ropes, used to support men working on the yard

g. Lifts of the main yard

i. Reef tackles of the topsail

l, m, n, Reef points of the topsail

(From Falconer's Marine Dictionary*)*

corners of the sails were controlled by means of the 'sheets'. The sheets could serve two different functions, according to whether they were attached to a course or an upper sail. In the latter case, the sheet simply kept the sail tight to the yard below. It passed through a sheave in the lower yardarm, through a block near the centre of the lower yard and then down to the deck below. The angle of the sail was controlled by the braces, and little adjustment was made to the sheets except when the sail was to be reefed or furled.

On the main and fore course, the sheets were the main means of controlling the angle of the lower part of the sail to the wind. Such a sheet had one end attached to a ringbolt on the outside of the hull, some way aft of the sail, and then passed through a block in the corner, or 'clew' of the sail, and then back through a sheave set in the side of the ship, near the position of the ringbolt. The sheet only held the clew from astern; forward from the clew was led the 'tack', to hold it forward, especially when sailing close to the wind. The tack of the foresail led to the boomkin projecting from the knight heads, while that of the main course led to a sheave in the side of the ship. It was also necessary to hold the edges of the sails forward when sailing close to the wind. This was done by means of 'bowlines'. Each sail had two or three cringles in its sides, and to these were attached the 'bridles' of the bowlines. These met a few yards forward of the sail, and the bowline itself led forward to a block set in an appropriate place; the deck for the main course, the foremast for the main topsail and topgallant, and the bowsprit for the foresails.

Courses were furled by hoisting the lower edge and corners of the sail up to the yard. For this purpose they were equipped with ropes known as 'buntlines', leading from the lower edge of the sail; and with 'clewlines', leading to the corners. Both types of line led to blocks slung under the yardarm above, and then down to the deck. The furling of upper yards was similar, except that the yard above was dropped a little first to take the wind out of the sail. They too were provided with buntlines and clewlines. Reef tackles were attached to the reef cringles at the ends of the lines of reef points which served to stretch the upper edge of the sail when it was reefed.

The fore and aft sails had slightly different systems of rigging. The mizzen course had 'brails' which served the function of buntlines and clewlines on a square sail. Its lower, forward corner had no need to move, and was held in place by a tack leading to the deck. Its sheet was attached to the lower, after corner, and controlled the movement of that part. The upper corner of its yard was controlled by ropes known as 'vangs', leading from the upper, aftermost end of the yard to the outer corners, or quarters, of the taffrail.

The staysails and jibs were different again. Each had a 'halyard', to haul it up the appropriate stay. It also had a 'downhaul' which was led in the opposite direction and was used for hauling it down. Such sails had no need of buntlines or clewlines, but each had two sheets on its lower, aftermost corner. In addition staysails, with their four corners, needed tacks to hold the forward, lower corner in place.

14

Fittings

Steering

The ship was steered by means of a rudder hinged on the sternpost which was fitted before the launch. It was vulnerable when the ship hit the water for the first time, but this was outweighed by the difficulties encountered in fitting it after the ship was afloat. The rudder was more than 30ft long, reaching from the base of the false keel to just above the level of the upper deck. Its width was equal to that of the sternpost, and there was no attempt to taper it fore and aft. It was comparatively shallow in its fore and aft dimension, being at its greatest point, at the false keel, only 6ft 3in deep. As it rose it became narrower, first gently, then more drastically by means of three steps, at the level of the wing transom, the gundeck and about halfway between the two. As a result, the rudder was almost square in cross section near its head and here it was cut with two square holes for the fitting of the tillers: one just above and one just below the level of the upper deck. The latter was the main fitting, and the other was only used in emergency.

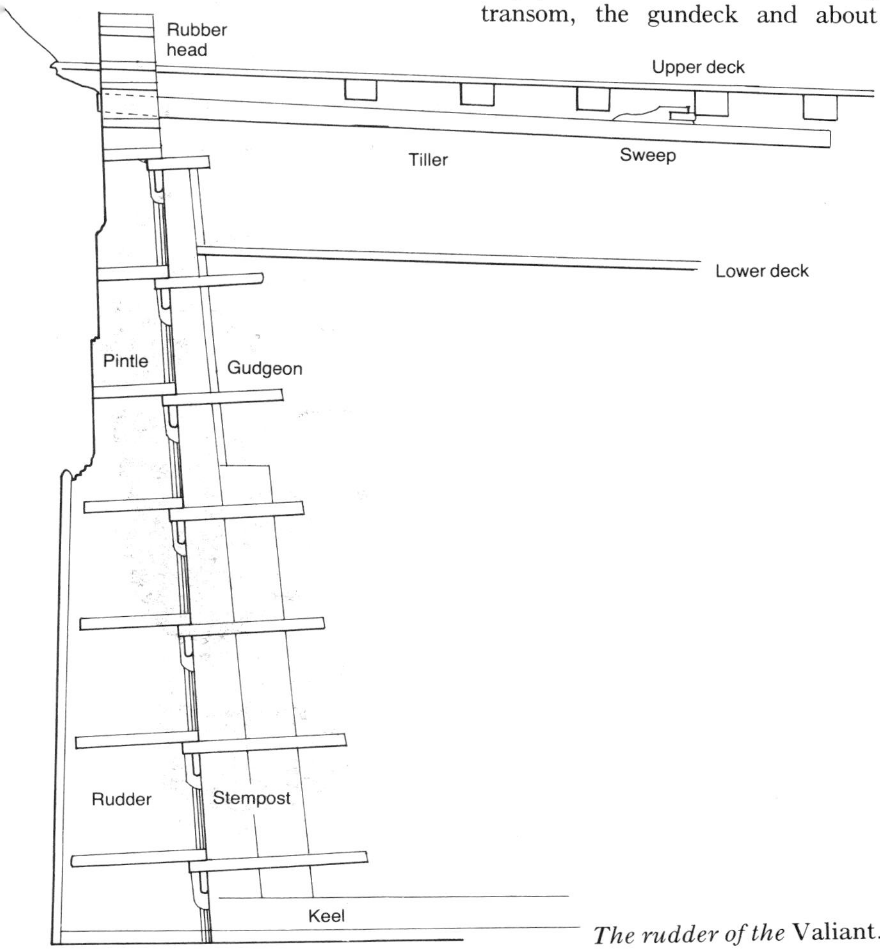

The rudder of the Valiant.

The forward edge of the rudder was straight, and in cross section it was slightly rounded to allow it to pivot on the sternpost. This edge was also cut with recesses for the 'pintles' which formed the upper parts of the hinges of the rudder. Seven of these were fitted, spread equally along the length. Each had two long arms by which it could be securely bolted on each side of the rudder. They were matched by 'gudgeons' on the sternpost, with arms spreading over the planking of the stern and with holes to receive the pintles. Wedges were used to prevent the rudder from floating up and out of the gudgeon.

The tiller was simply a straight piece of timber fitted to the square hole in the rudder below the upper deck. Its forward end was supported by a 'sweep' – a curved piece of timber under the upper deck beams. The tiller lines led from each side of the fore end of the tiller, through blocks on each side of the hull, back to sheaves set in the deck just forward of the tiller end in its neutral position, and then up to the steering wheel. The line was led several times (traditionally seven) round the barrel of the wheel, nailed to it in the middle, and then led back to the deck below.

The wheel of Victory, with the binnacle in front of it. (Author's photograph, courtesy of HMS *Victory)*

The wheel of a ship of the line was double; it had two steering wheels, of the traditional nautical form with spokes projecting from its rim, one set at each end of a cylindrical 'barrel' and this allowed up to four men to handle the wheel in heavy weather. It was placed under the forward part of the poop deck, where the helmsmen could be sheltered but still see what was going on. Each end of the barrel was supported by a decorated pillar, and the forward one went all the way up to join the foremost beam of the poop deck.

The 'binnacle' was a box holding two compasses, with a candle between them for use at night, and this was placed just forward of the wheel where the helmsmen could see it. A kind of helm indicator was often fitted above the wheel, on the forward side of the poop deck beam. It was operated by lines from the wheel which moved an arrow along a large scale on the beam. Such an indicator had been found aboard the *Invincible* on her capture in 1747 and was ordered to be copied for British ships. It allowed the officer of the watch to see what angle the rudder was set at, and therefore to judge the performance of the ship.

Pumps

Wooden ships always leak no matter how well constructed and caulked they are. Water could enter through the sides of the ship or be shipped

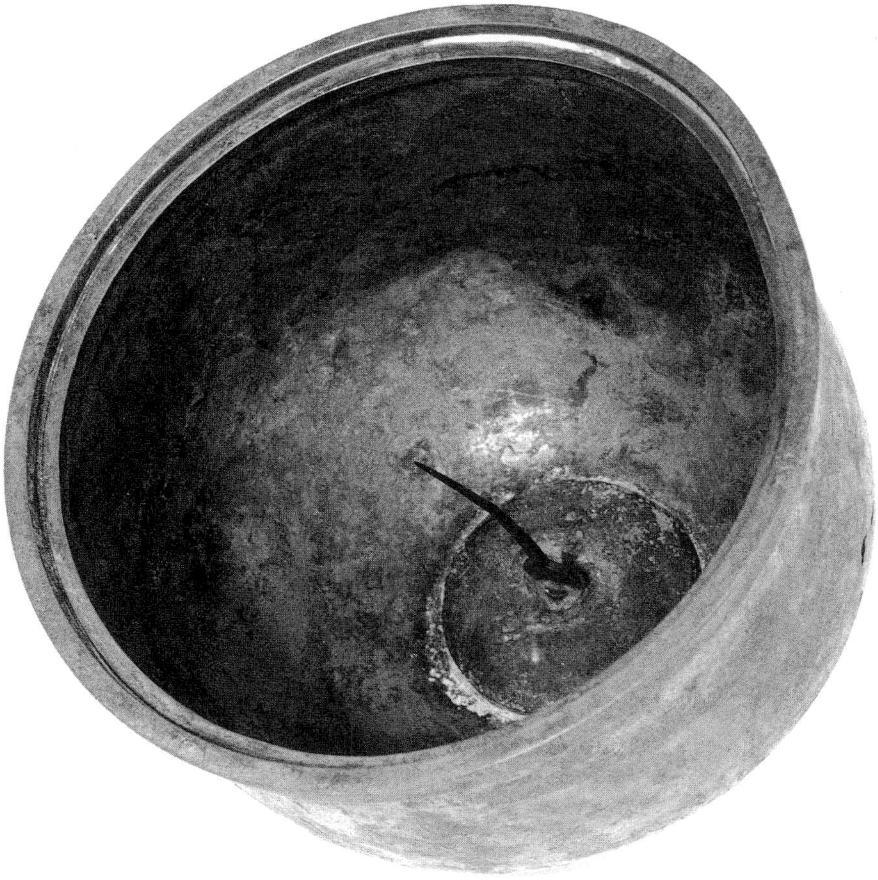

A compass bowl recovered from the wreck of the Invincible, *showing the pin which supported the card and which allowed it to rotate.*

over the side, or rainwater could find its way through the decks; a system of drainage was an intergral part of the design of any vessel. The camber of the decks allowed water to run to the sides. The planks of the sides were protected by the waterways, and the coaming prevented it from falling down the hatchways. The 'scuppers' allowed the water to drain out. These were lead pipes, set at intervals in holes bored diagonally through the waterway and the side of the ship. The foremost part of the lower deck was particularly prone to receive water through the hawse holes, and that area was sealed off by a light partition, known as the 'manger', which did not go up to the full height of the deck above; a scupper was provided on each side in the area enclosed by the manger. Below decks, the water in the hold was encouraged to drain to the centre of the ship, around the mainmast. This area was known as the 'pump well', and was enclosed by a light partition to keep the ballast and cargo away from it.

The ship had four main pumps, arranged round the mainmast. The lower end of each was set just above the floor of the ship, and it discharged above the lower deck, as this was the first deck above the waterline. Chain pumps had been used on British ships since Elizabethan times. Each consisted of a tube, with a chain running up it, and saucers fixed at intervals to the chain, with leather discs to make the seal tight. The chain was endless, and ran over a wheel which was set with brackets to hold the saucers. This wheel was placed above the lower deck, and cranks could be passed through so that seamen could turn it. After passing over the wheel the chain went down again, guided by another tube, in which it fitted less tightly. At upper deck level, the pump discharged into a wooden cistern and then into a wooden pipe known as a 'dale' which led to the side of the ship. There the water was discharged through the scuppers.

Anchors

A ship like the *Valiant* needed six anchors. The four largest ones were known as the best and small bower, the sheet anchor and the spare. These were approximately equal in size and, on a 74-gun ship, weighed about 70cwt each. The stream anchor was much smaller, about 16cwt, and the kedge was the smallest of all, weighing about 8cwt. This kedge was designed to be slung under a boat and rowed out ahead of the ship, as a way of moving it in confined waters. The bower anchors, as their name suggests, were stowed in the bow, suspended from the catheads. The sheet and spare anchors were stowed just behind them, with one lashed against the fore channels on each side. The stream and kedge anchors were lashed to them.

An anchor had a long straight piece, known as the 'shank'. A large ring was fitted to one end of the shank, and used to attach the anchor cable. Just below this was the 'stock'. This was made of wood, with two halves being joined together by iron bands. At the other end of the shank were the 'arms' of the anchor. These projected at an angle to one another, set in a plane at right angles to the stock. Where they met was known as the 'crown' while the ends were fitted with 'flukes', spade-like pieces, one of which would bury itself in the sand or mud and provide most of the holding power

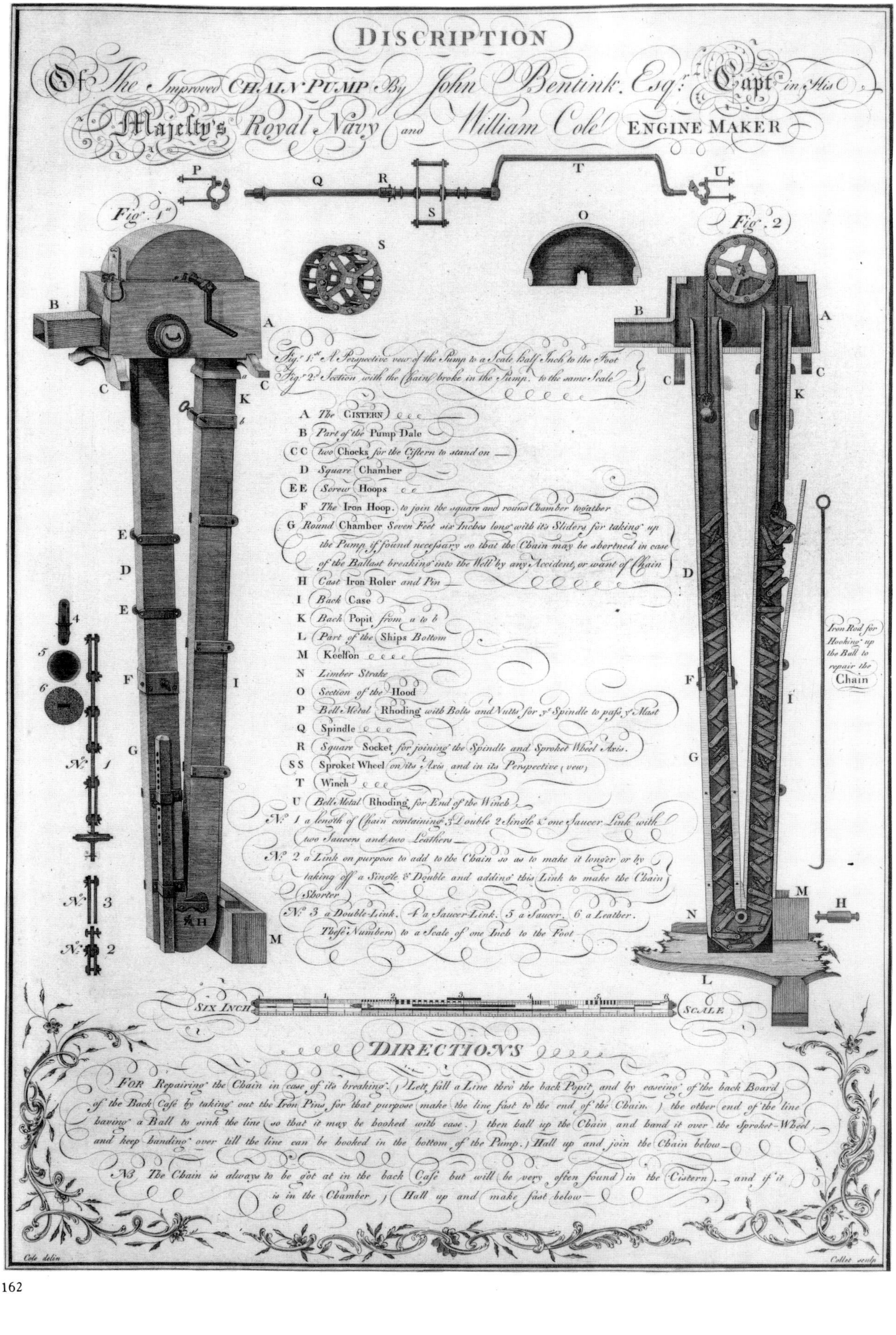
DISCRIPTION
Of The Improved CHAIN PUMP By John Bentink. Esqr. Captn. in His Majesty's Royal Navy and William Cole ENGINE MAKER
P
Q
R
T
U
S
Fig. 1st.
O
Fig. 2
S
B
A
B
A
C
C
C
C
K
K
a
b
Fig. 1st. A Perspective vew of the Pump to a Scale Half Inch to the Foot
Fig. 2d. Section with the Chain broke in the Pump. to the same Scale
A The Cistern
B Part of the Pump Dale
CC two Chocks for the Cistern to stand on
D Square Chamber
EE Screw Hoops
F The Iron Hoop. to join the square and round Chamber togather
G Round Chamber Seven Feet six Inches long with its Sliders for taking up the Pump if found necessary so that the Chain may be shortned in case of the Ballast breaking into the Well by any Accident, or want of Chain
H Cast Iron Roler and Pin
I Back Case
K Back Popit from a to b
L Part of the Ships Bottom
M Keelson
N Limber Strake
O Section of the Hood
P Bell Metal Rhoding with Bolts and Nutts for ye Spindle to pass ye Mast
Q Spindle
R Square Socket for joining the Spindle and Sproket Wheel Axis.
SS Sproket Wheel on its Axis and in its Perspective vew
T Winch
U Bell Metal Rhoding for End of the Winch
No. 1 a length of Chain containing 3 Double 2 Single & one Saucer Link with two Saucers and two Leathers
No. 2 a Link on purpose to add to the Chain so as to make it longer or by taking off a Single & Double and adding this Link to make the Chain Shorter
No. 3 a Double Link. 4 a Saucer Link. 5 a Saucer. 6 a Leather.
These Numbers to a Scale of one Inch to the Foot
E
D
E
F
I
G
H
M
4
5
6
No. 1
No. 3
No. 2
D
F
I
G
M
H
N
L
Iron Rod for Hooking up the Ball to repair the Chain
SIX INCH SCALE
1
2
3
4
5
6
DIRECTIONS
FOR Repairing the Chain in case of its breaking. Lett fall a Line thro the back Popit, and by easeing of the back Board of the Back Case by taking out the Iron Pins for that purpose make the line fast to the end of the Chain. the other end of the line having a Ball to sink the line (so that it may be hooked with ease.) then hall up the Chain and hand it over the Sproket Wheel and keep handing over till the line can be hooked in the bottom of the Pump.) Hall up and join the Chain below
NB The Chain is always to be got at in the back Case but will be very often found in the Cistern. and if it is in the Chamber, Hall up and make fast below
Cole delin
Collet sculp

◄ *Details of a chain pump, as improved by Coles and Bentinck in the 1760s and 70s. This print was issued to the carpenters of ships to aid them in its maintenance. (National Maritime Museum)*

of the anchor.

Anchors were made in the Royal Dockyards, by forging: heating small pieces of metal and then hammering them until they formed a solid mass. The forge at Chatham was visited by an anonymous diarist in 1759, during the building of the *Valiant*. He noted, 'we had the pleasure of being in the smith's shop at the instant the several workmen were turning an anchor of a man of war of four ton weight then in the fire; all our attentions were engrossed as it was a surprising thing, and we were filled with horror at the glowing heat of the several workmen, who with great dexterity managed the affair; anchors are made by hammering piece upon piece'.[1]

Smiths worked in small teams, each under a foreman. Chatham Dockyard had seventy-four smiths in 1759, working in a smithery sited between the dry docks. It was replaced by a new building in 1808, and this still stands today. The smithery had a number of fires for heating the iron, served with bellows. Various types of crane were used for lifting the anchors, according to their size. The beating could be done by a single hammer in the case of a small anchor, or by an instrument known as a 'hercules' for the largest ones.

Capstans

The ship was fitted with two capstans, made of wood by the shipwrights. They were principally designed for raising the anchors but were also needed for lifting in the guns, and raising the heavier spars when fitting out the ship. The main capstan was placed halfway between the mainmast and the mizzen, under the quarterdeck. The 'fore jeer capstan' stood in the waist, about halfway between the foremast and the mainmast and was identical in size. The core of each capstan was the 'barrel' which rested on the lower deck with its upper part well above the upper deck, and in the

The anchors of a 74-gun ship. (National Maritime Museum)

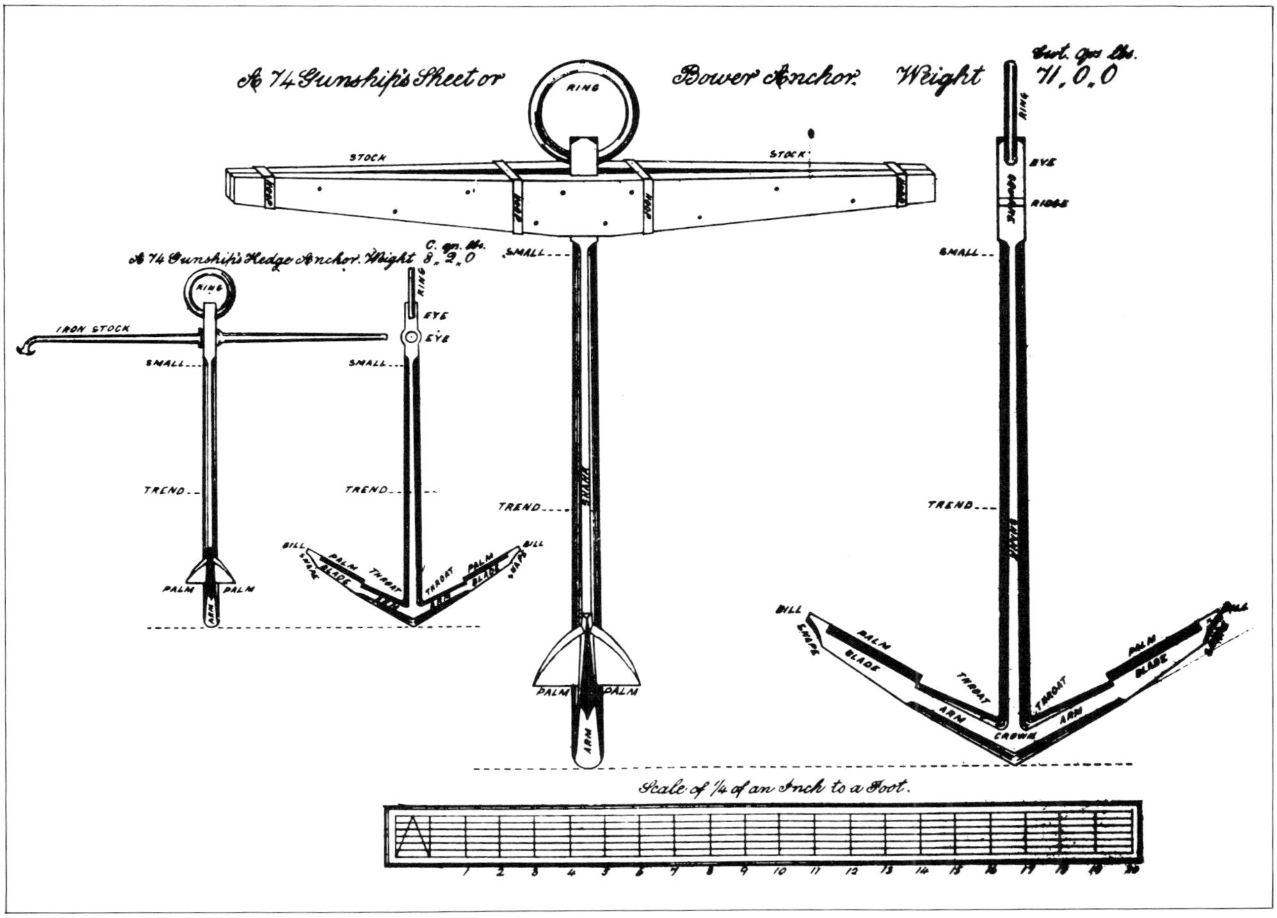

case of the main capstan, almost reaching up to the beams of the quarterdeck. The barrel was fitted with 'whelps' above both lower and upper deck level. These were irregular shaped pieces of timber which projected out from the barrel and served to both extend the radius round which the cable could be coiled, and prevent it from riding up. Above the whelps were the 'heads' of the capstan, one for the upper deck and another known as the 'trundle head', for the lower deck. Capstans of this period had flat circular heads known as 'drumheads' and these had a dozen squarer holes round their rims into which were inserted the capstan bars which the men pushed to turn the capstan. For very heavy work the two heads could be used at once and the two capstans could be used in tandem employing nearly 400 men.

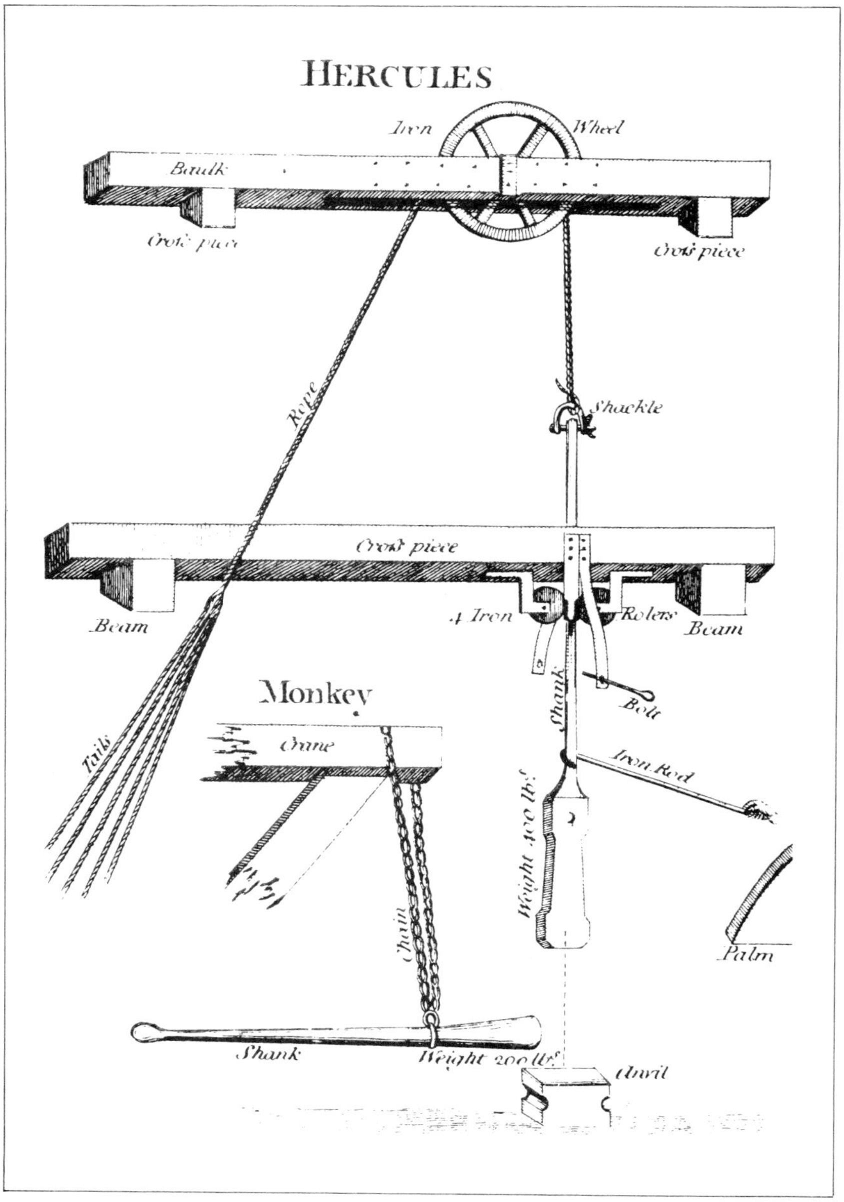

Anchor making equipment, including a Hercules hammer. (*From Steel's* Rigging and Seamanship)

Anchor Tackle

The anchor cables of the *Valiant* consisted of rope of 21in circumference and about 120 fathoms, or 720ft, long. Such cables demanded the full length of the ropeworks in their making. Aboard ship they were stowed in the middle part of the orlop deck in large coils. Eight or nine were issued so that more than one anchor could be used at once, or several cables joined together if the ship was bringing up in deep water. In addition, lighter cables were issued for the stream and kedge anchors. The main cables were too thick to turn directly round the capstan and so an endless rope known as a 'messenger' passed round the capstan itself and then led forward to meet the cable coming in through the hawse hole. It was fixed to the cable by means of light ropes known as 'nippers' and as the cable was hauled in the nippers were taken off in turn and the messenger passed forward to meet the cable further along. The cable was passed down through the main hatch, to be stowed by a large party working on the orlop deck.

After the anchor had broken surface it was important to keep it clear of the side of the ship. The cathead served as a kind of crane and a tackle, led through a large block known as the cat block, was used to lift the anchor clear of the water. Once this was done, the crown of the anchor was also raised so that the shank became horizontal. The 'fish tackle' was used for this and was suspended from the 'spanshackle boom', a wooden beam which could be moved from side to side over the forecastle, and whose inboard end was held by a large shackle in the middle of the forecastle. The fish tackle included a

A model of a typical capstan of the mid-eighteenth century, made from timber from the Royal George *of 1756.*
(Author's photograph)

A model of a longboat, cutter-rigged and carrying oars and a grapnel.
(Author's photograph, courtesy of the National Maritime Museum)

A LONG-BOAT FOR A THIRD-RATE

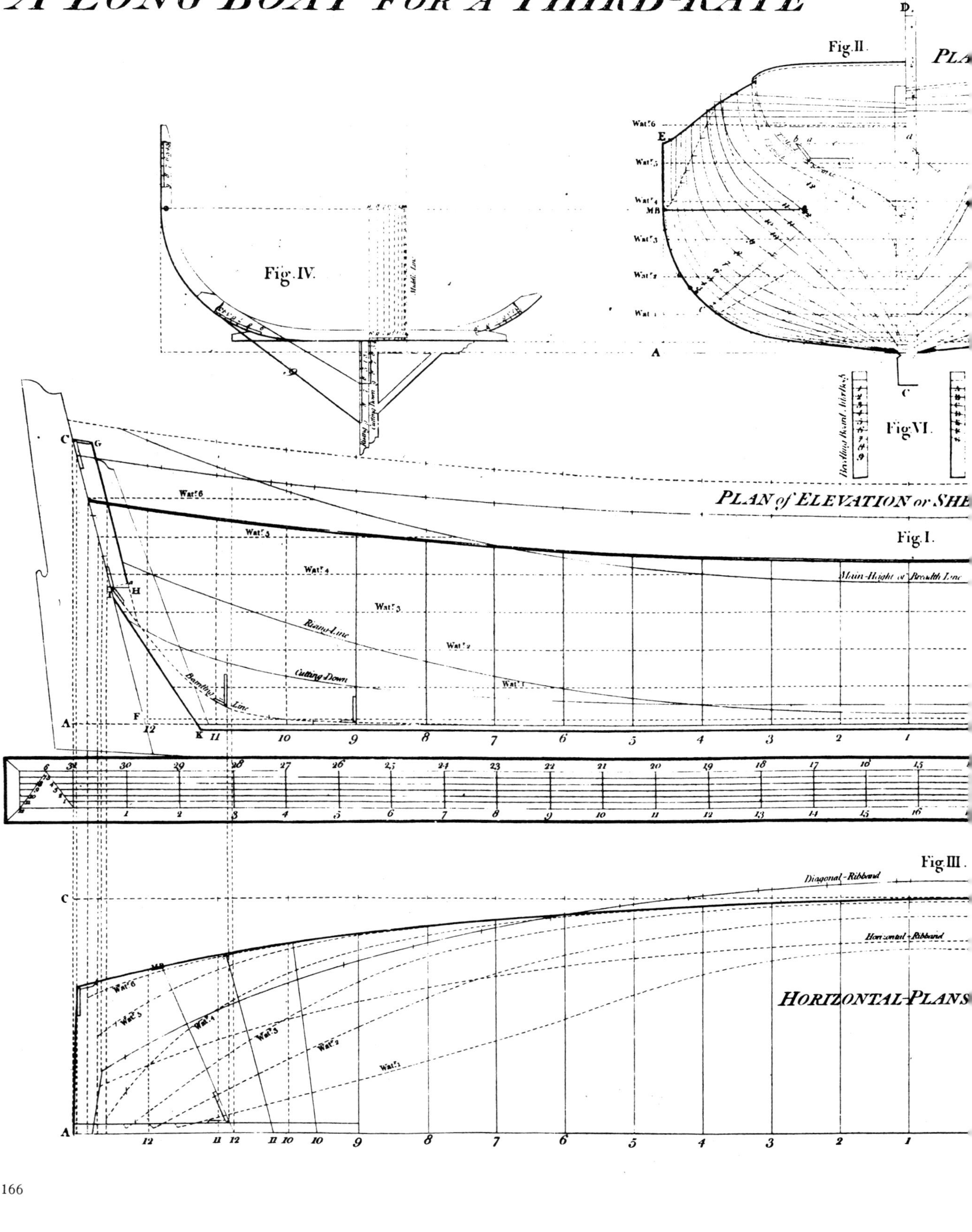

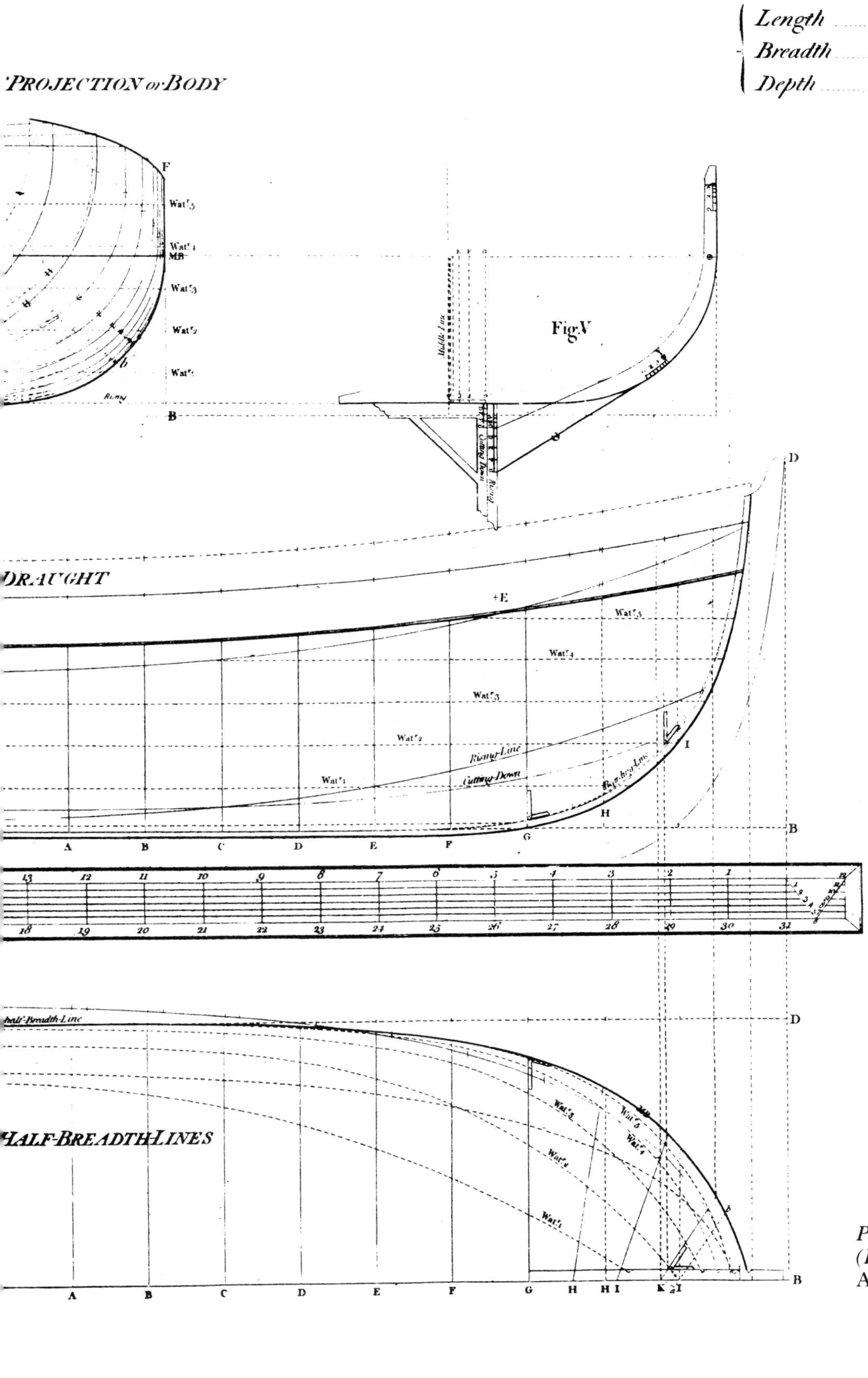

Plans of a longboat. (From Stalkaart's Naval Architecture *of 1782)*

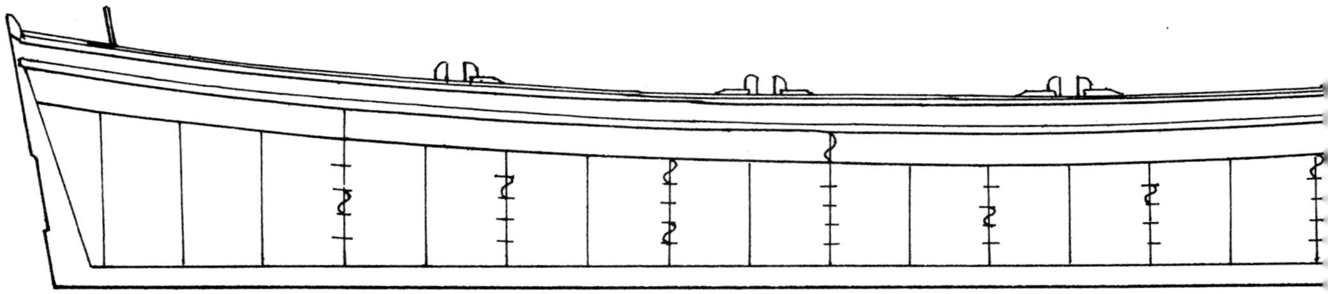

A pinnace of 1795. The lines of the sides indicate the run of the planks and the 'S' shapes show the joins.

A cutter, as carried by Captain *in 1762.*

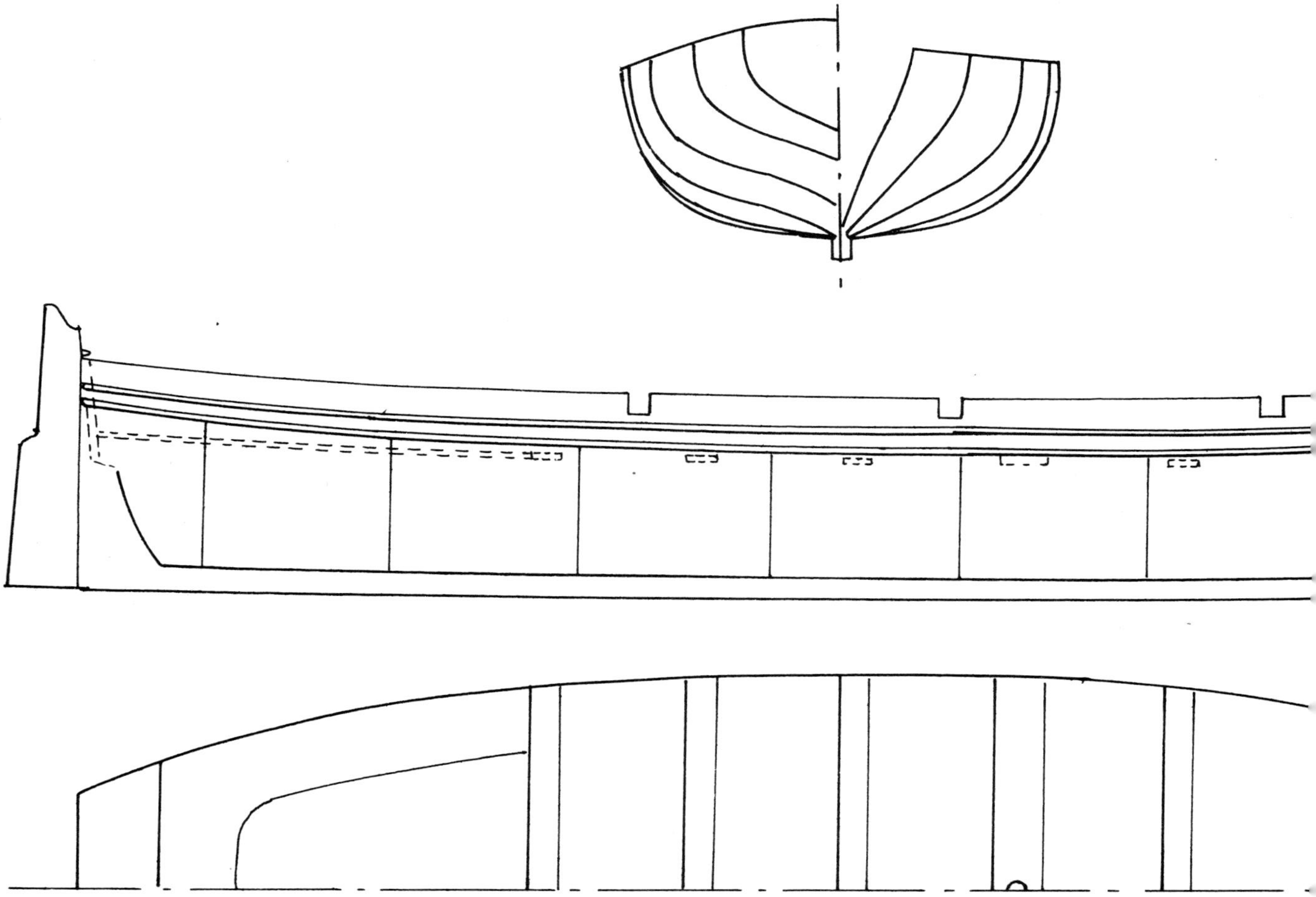

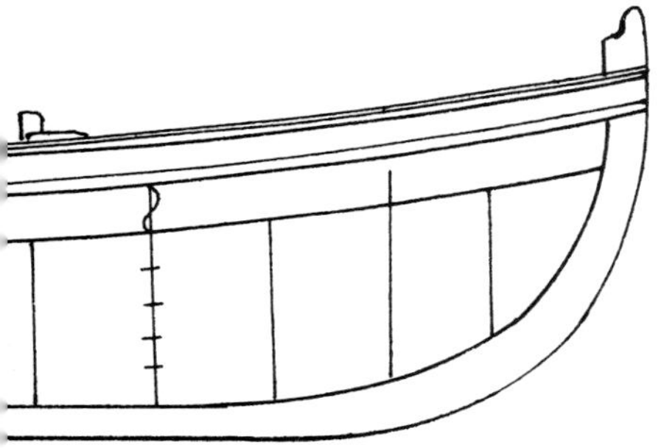

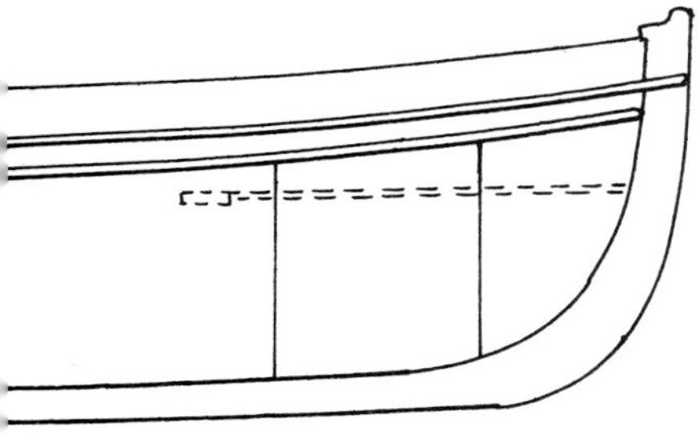

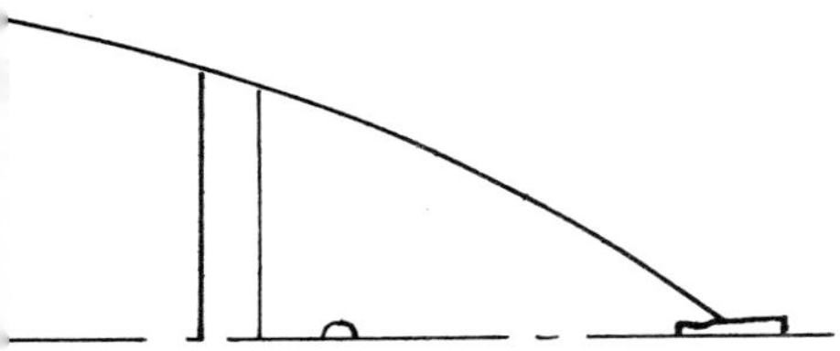

large hook which could be attached to the shank of the anchor. After being raised in position, the anchor was lashed up by means of ropes which could be released quickly, and the cat and fish tackles were then taken off. At sea, the cable was removed from the ring of the anchor, and the hawse holes blocked up to prevent the entry of water.

Boats

A ship like the *Valiant* carried five or six boats, of different designs and serving many purposes. The longboat was 31ft x 10ft and used for heavy work, such as moving anchors and carrying stores. Despite the name, it was not the longest boat carried. This was one of the two 'pinnaces' which was 32ft x 6ft 10in, and being long and narrow was good for rowing in harbour though less efficient under sail or at sea; it was the captain's personal boat and used to take him ashore or to another ship. The smaller pinnace was 28ft x 6ft 6in and was used by the junior officers, particularly the lieutenants. Two cutters were carried, one 25ft x 7ft 4in, the other 18ft x 6ft 2in. These were proportionately broader than the other boats and with their sharp bows were good sea boats and efficient under sail. Last, she carried a yawl, 25ft x 6ft 4in. This boat was similar to the large cutter, but with slightly blunter lines and used for much the same purposes.

These boats were either carvel- or clinker-built. The longboats and pinnaces were carvel which was the method familiar to the shipwrights of the Royal Dockyards, and therefore these boats were mostly made there. In contrast to the 'frame first' system used in carvel building, clinker boats were built 'skin first'. Each plank was carefully shaped and then fixed to the one below by means of nails or 'clenches', with their heads turned over to secure them. A light frame was added later to give additional strength, but the real strength was from the planks themselves. Clinker build has advantages and disadvantages. The joins between the planks gave extra strength against bending forces, but on the other hand it was difficult to replace defective planks without dismantling the whole structure. The planking has to be kept thin, and this limits the possibilities of building large boats. Naval shipwrights had no training in this type of work and the clinker boats, comprising cutters and most of the yawls, were built in boatyards in the area around Deal in Kent where there was a long tradition of clinker boat building.

Boat Fittings

All boats could be sailed or rowed, but some were better at one than the other. Pinnaces were essentially fast rowing boats, and could be rowed 'double banked', with a rower sitting on each side of the thwart; a 32ft pinnace could be rowed by ten men. Cutters were generally 'single banked' and the rower sat on the opposite side of the thwart from where the oar projected. This produced extra leverage but halved the number of oars that could be fitted. An 18ft cutter was usually fitted for four oars.

Boats were invariably fore and aft rigged, with one to three masts. Longboats were generally cutter rigged, with a single mast carrying a gaff sail, and a staysail and a jib set from the bowsprit. Cutters of this period had a foremast and a mizzen mast, and were 'spritsail rigged'. (A spar known as a 'sprit' projected up and outwards from the lower part of the mast to the upper corner of the sail.) Cutters could also carry gaff, lug and lateen sails, and usually had two or three masts.

Boats needed a whole range of equipment to keep then operational. Each was fitted with a rudder and 'grapnel' anchors were carried, with four arms instead of two so that a stock was unnecessary. Longboats had heavy lifting gear in the form of a windlass and davit and all required boathooks for coming alongside and for picking up objects from the water.

Most boats were kept in the waist of the ship, on booms placed between the gangways joining the forecastle and quarterdeck. These were quite a recent innovation, and were removable. The boat was kept upright by wedges and shores, or by a special stand with its upper part cut to the section of the part of the boat that was to rest in it. The boats were lashed down, using eye-bolts placed in the booms and protected by a canvas cover.

At this period there were no specially-designed davits for raising and lowering boats, and a system of tackles was used, similar to that for moving guns and heavy stores. A tackle was suspended between the tops of the foremast and mainmast, and two more tackles were hung from that, one for each end of the boat. Two further tackles were needed, from the ends of the fore and main yardarms. The weight was taken up by the first tackles, and then transferred to the yardarm tackles so that the boat swung out over the water from where it could be lowered into the sea.

The boats of a ship of the line gave it a great deal of flexibility in service and were used for providing propulsion for the ship in a calm, bringing aboard all kinds of stores, landing troops and for communication between ships or with the shore.

15

Guns

The guns were the *raison d'etre* of a ship of the line, and her principal means of offensive action. The guns and their carriages, ammunition and associated tackle, were the only parts of the ship which were not supplied and controlled by the Admiralty and its subsidiary boards. Armaments for both the army and navy were supplied by the Ordnance Board, a separate government department. In the seventeenth century there had been many disputes between the Admiralty and the Ordnance Board, and as recently as 1733 the Ordnance Board had delayed changes in ship armament because it did not have stocks of the new types of gun needed; but by 1759 the Ordnance Board was doing better at arming the new 74-gun ships as they came off the stocks.

Gun Making

A ship's cannon, or 'great gun', was simply a strong iron tube, sealed at one end except for a small hole which was used to ignite the powder. In earlier times some guns had been made of brass while many iron guns had been made up piece by piece but by the mid-eighteenth century brass guns were extremely rare in the British fleet and most guns were cast in a single piece, made in a mould. They were smooth bored, and muzzle loading. The government had no facilities of its own for making iron cannons, and they were made by private firms, under contract. In 1759 the

Gunmaking – the mould in use. (From the French Encyclopedie *of 1751–66)*

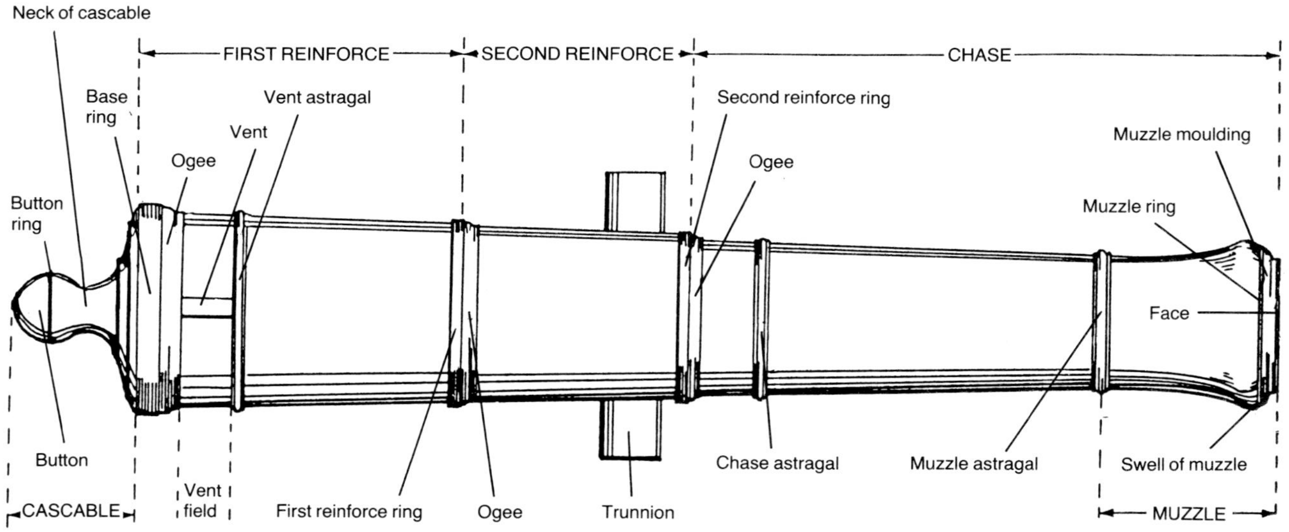

The parts of a gun.

Weald of Kent and Sussex was still the premier gun-founding district of the country, but it was about to lose its position to Scotland and the North of England. It was still common to cast guns using charcoal but new processes, using coke, were being developed and the Weald was soon to decline.

The process of gunmaking began with the forming of a full-size model of the gun, including a large 'head' which would be cut off after manufacture. The model was used to make a female mould, usually of sand at this period. A cylindrical piece of wood was inserted to form the bore of the gun and the whole assembly was turned to point upwards. Molten iron was poured in and allowed to cool. The mould was broken open and the gun was ready. It was sent to the Ordnance depot at Woolwich, where it was subjected to proof. It was buried in sand and filled with a large amount of powder, more than its normal charge. It was fired and then examined minutely for any signs of damage. If it failed the trunnions were broken off to make it useless; if it passed it was sent to one of the Ordnance Board depots, perhaps the Ordnance Wharf at Chatham.

The length of a gun was measured from its front face to its thickest part at the 'base ring'. This left the rearmost part of the gun, the 'cascable', which included a substantial spherical part known as the 'button' to which were attached the ropes which restrained the gun when fired. Forward of the base ring, the guns tapered by means of a decorated ogee, and then there was a short section known as the 'vent field', containing the touch hole or 'vent'. This formed part of a larger area known as the 'first reinforce', amounting to about a quarter of the length of the gun. This was the thickest part of the gun for thereafter it tapered again to the 'second reinforce ring'. The trunnions were fitted just behind this ring and projected sideways from the gun, a little below its centreline. They formed a pivot for the elevation of the gun, and provided a secure means of fixing it to its carriage.

Forward of the second reinforce ring, the diameter of the gun narrowed sharply. It then resumed its gradual taper, passing the decorative 'muzzle astragal', until it reached the 'neck' where it was at its narrowest; there it bulged outwards, to form the 'swell of the muzzle', followed by a sudden reduction in diameter, and a series of decorative mouldings leading to the flat face of the muzzle. The whole area forward of the second reinforce ring was known as the 'chase'.

Types of Guns

The 32-pounder gun, firing a solid iron ball of that weight, formed the main armament of the *Valiant*. In a series of experiments conducted in 1747, the mathematician Benjamin Robins had established that the 32-pounder was the most efficient gun in service. It had longer range and did greater damage to the hulls of enemy ships, in proportion to its weight, then smaller weapons. The only larger gun in the navy was the 42-pounder which was not tested by Robins. This was the main armament of the three-deckers of the First Rate, but its ball was too heavy for a man to handle in action. The 32-pounder was the main armament of the new 74-gun ships which were being built, as well as the three-decker of 90 guns. It soon proved to be the most efficient type of heavy gun.

On her upper deck the *Valiant*, like her sister the *Triumph*, carried 24-pounders. This made her unusual among two-deckers, for the great majority carried 18-pounders. The *Invincible* had been fitted with the larger guns in 1757, though she never used them in action. The 24-pounders of the *Valiant* made her a very powerful ship indeed but they were not a success. In 1778 Admiral Keppel, who probably knew the ship better than anyone else, wrote, 'I should hope that she will be ordered 18-pounders; and if after that you like to copper her, she will outsail the coppered frigates and be the completest ship in the world'.[1] Evidently, the 24-pounders were too heavy for the ship, and they were replaced soon afterwards.

On her quarterdeck and forecastle the *Valiant* carried 9-pounders in common with most ships of the line. Such guns made only a slight contribution to the gun power of the ship, but their inclusion was justified by the argument that they were situated high up and so could fire down onto the decks of a smaller enemy, or into the rigging of a larger one. From 1779 onwards ships of the line began to carry carronades: shorter guns with a large calibre and so named because they were made in the Carron Iron Works in Scotland. After 1794 they began to replace some of the 9-pounders of the quarterdeck and forecastle. Carronades made a substantial contribution to the short range armament of a ship, but one effect of their introduction was that a ship no longer carried the number of guns implied in her rating; a 74 no longer carried precisely 74 guns, for example. The *Valiant*, however, had not been fitted with carronades by 1782, though she probably had some by the 1790s.[2]

Types of gun carried by the Valiant.
A 32-pounder

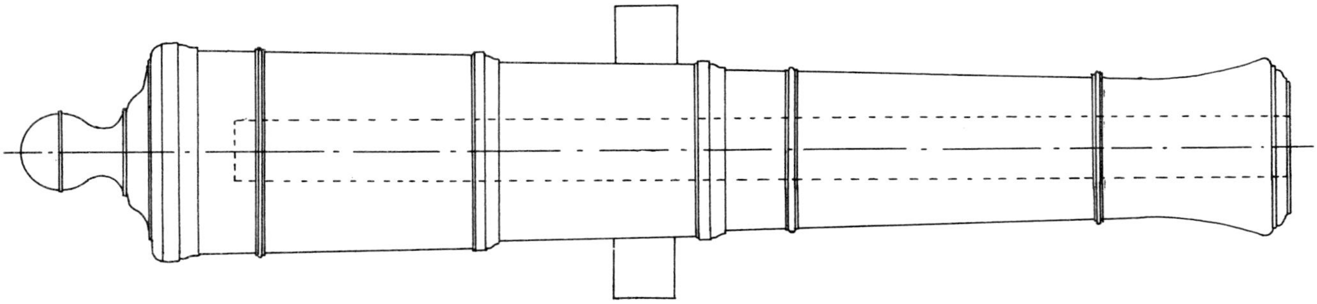

A 24-pounder

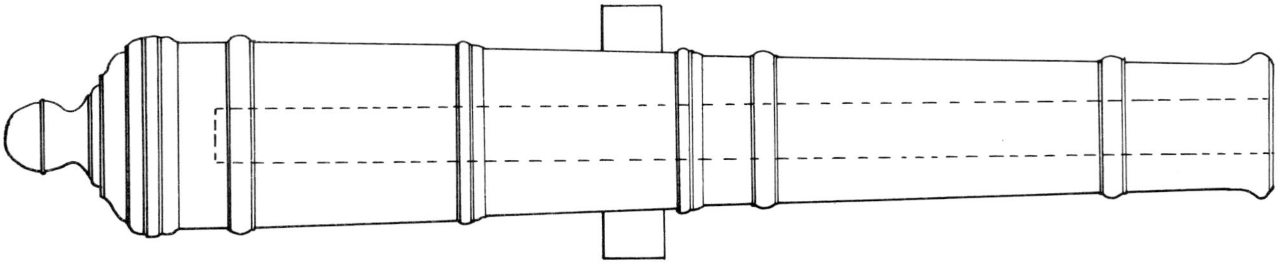

A 9-pounder

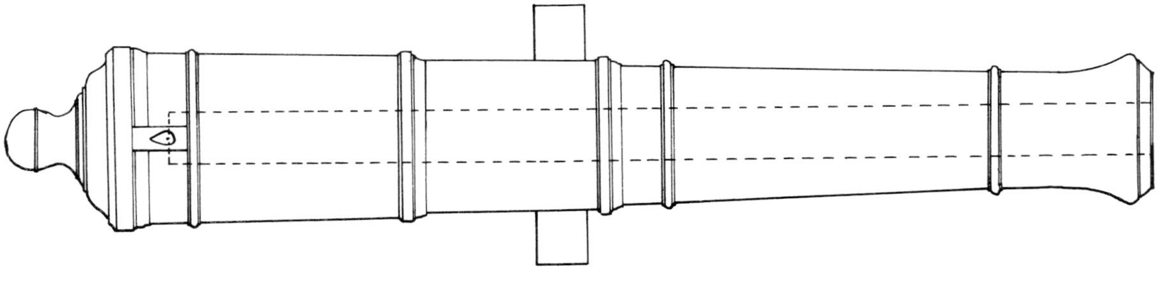

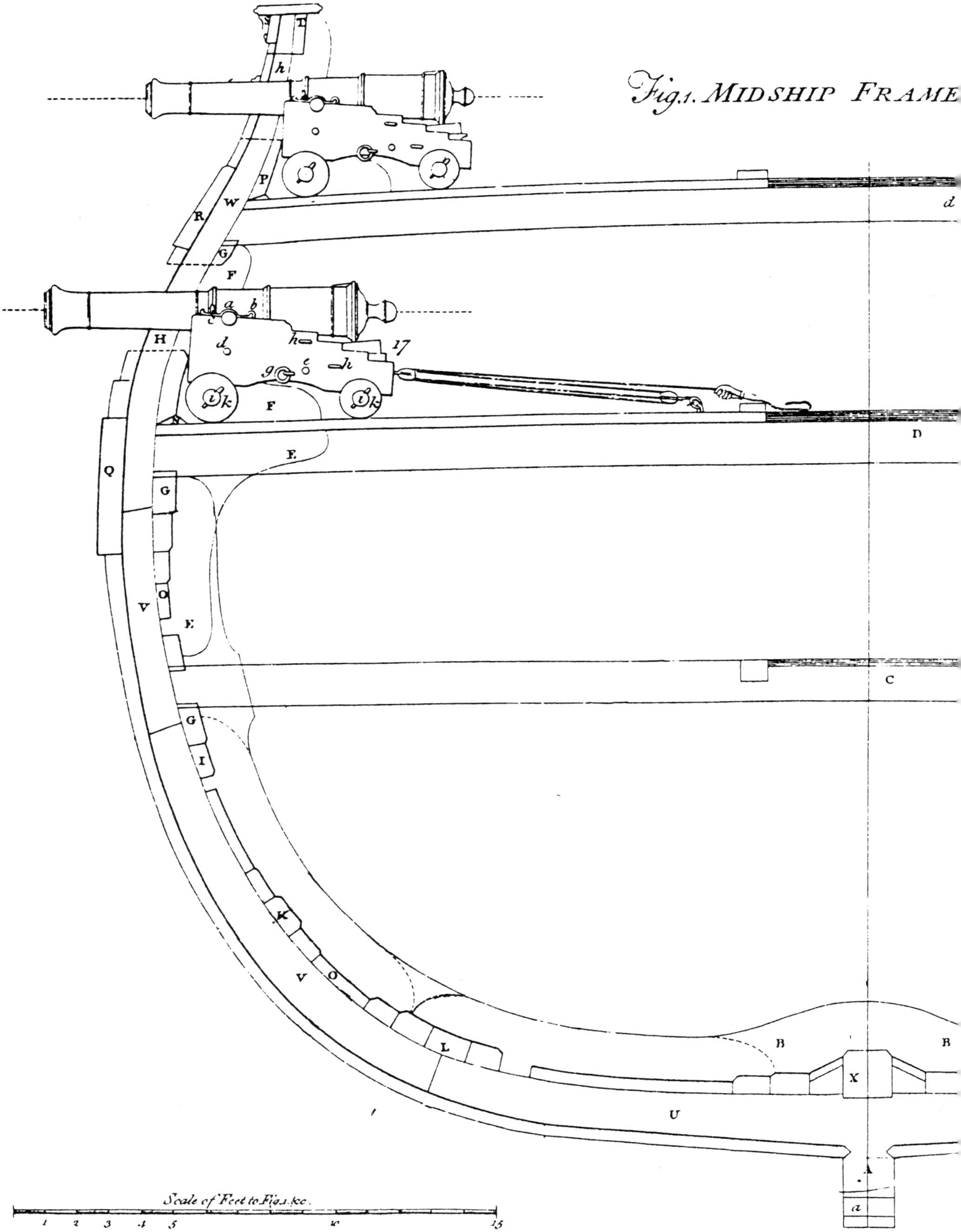
Fig. 1. MIDSHIP FRAME
Scale of Feet to Figs. &c.
1 2 3 4 5 10 15
S T h P W R G F a b c d h e g i k 17 H F E Q G V O E D C G I K V O L B B X U A a d

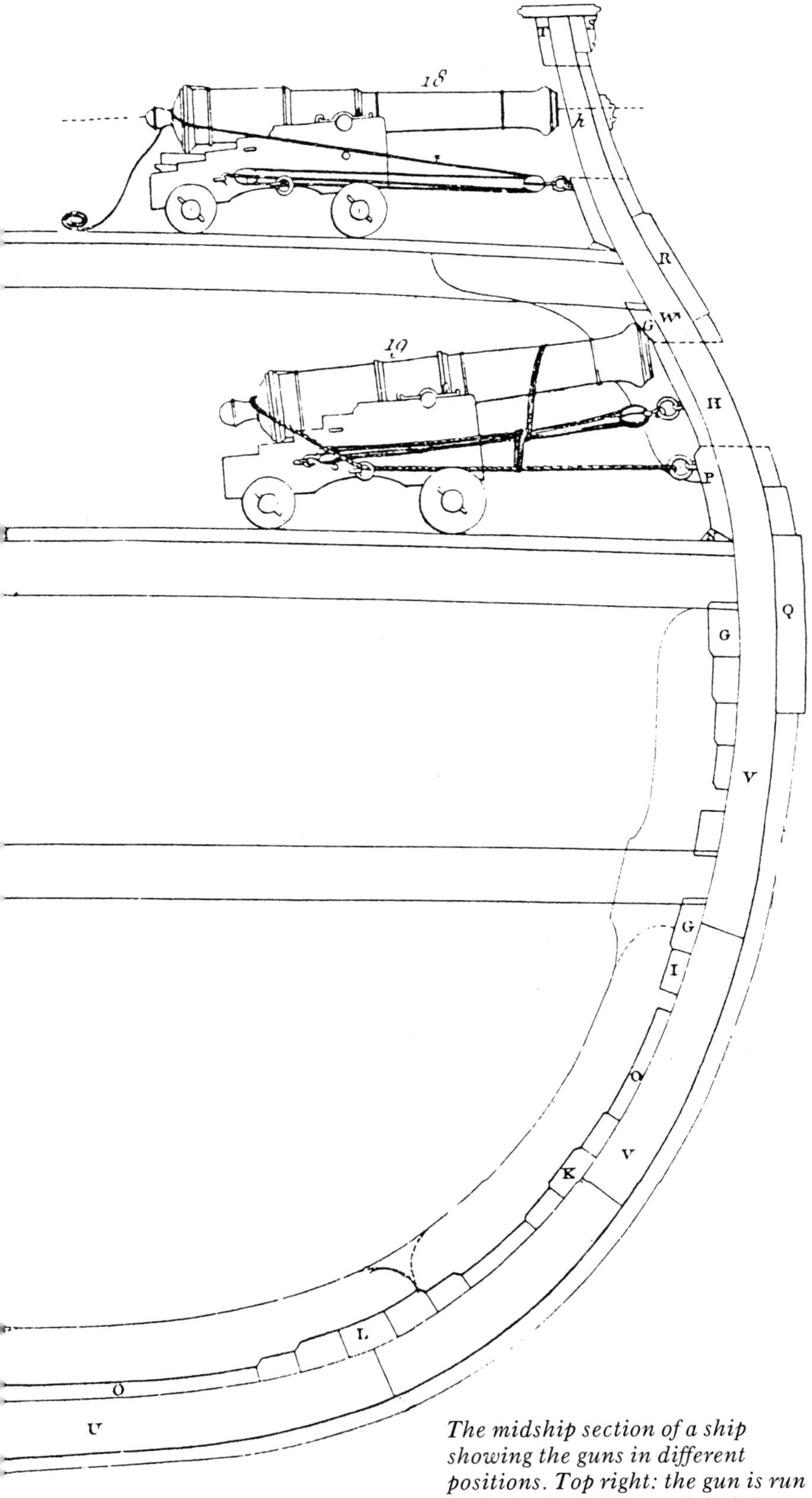

Carriages

Each gun was mounted on a wooden carriage fitted with four wooden wheels known as 'trucks'. The largest parts of the carriage were the sides, the 'cheeks'. Each was roughly rectangular in shape except that the after, uppermost corner was formed into a series of steps. Each cheek was made in two pieces, one above the other, and they were joined together by long bolts, passing all the way through their depth. Underneath, the cheeks were joined together by a pair of axles, which also served to hold the trucks. At a higher level, in the forward part of the carriage, the cheeks were joined by a solid piece of timber known as a 'transom'. A groove was cut in the upper, forward part of each cheek to receive the trunnion of the gun. These were held in place by curved metal plates known as 'capsquares' which could hinge upwards when required, and were held in place by the same bolts which held the parts of the cheeks together. The after part of the gun was supported by the 'stool bed', which rested on the rear axle. If necessary a wedge, known as a 'quoin', could be put in over this to help elevate or depress the gun. The sides of the beds were fitted with ringbolts to hold the tackles of the gun.

Carriages were also supplied by the Ordnance Board. They were made to standard designs for each size of gun, but the height had to be varied to match the height of the gunports on each individual ship and this information had to be supplied by the Admiralty for each new vessel building.

The midship section of a ship showing the guns in different positions. Top right: the gun is run in for loading. Bottom left: it is run out for firing. The gun and breeching tackles are omitted, but the train tackle is shown. Bottom right: the gun is housed for heavy weather. *(From Falconer's* Marine Dictionary*)*

A powder barrel from the wreck of the Invincible.

Gun Tackle

Clearly, a gun carriage running loose about the decks of a ship could have devastating effects, and several ropes were used to restrain each one. The thickest were the breech ropes, 7½in in circumference on a 32-pounder, and one was used for each gun. The middle part of the rope was turned round the button of the gun, and seized on to it; each end was then led to a ring bolt on the side of the ship and attached to it, probably with a bowline. This tackle was intended to restrain the recoil of the gun when it was fired. It was long enough to allow the gun to run some way back from the port and so be reloaded, but not so long that it ran onto the coamings and other fittings near the centreline of the ship.

In addition, each cannon had two gun tackles. These were of much lighter rope (3in on a 32-

Racking from the Invincible*'s magazine recovered from the wreck and reassembled.* ►

One of the powder mills at Faversham with the remains of others in the foreground.

Loading Guns on Board

The guns were put on board by the crew of the ship. The water at Chatham was shallow and a ship with its full complement of guns would be likely to go aground, so large ships were taken down to Blackstakes, near Sheerness, to have their guns fitted. In the case of the *Valiant* the process began on 10 October when a lighter arrived from the Gun Wharf carrying ten lower deck guns. The gun carriages were lifted on first and put on the appropriate decks. Tackles were rigged from the yardarms to lift the guns out of the lighter and another rope led from the muzzle of the gun through the gunport where it was to be situated. When the gun was level with the port it was hauled in. The gun was held level while the carriage was placed under it and then lowered. The operation continued on the 11th when the remaining lower deck guns were loaded on board along with the first eighteen of the 24-pounders for the upper deck. Work was interrupted by gales for the next two days, but by the 15th the ship was fully armed.

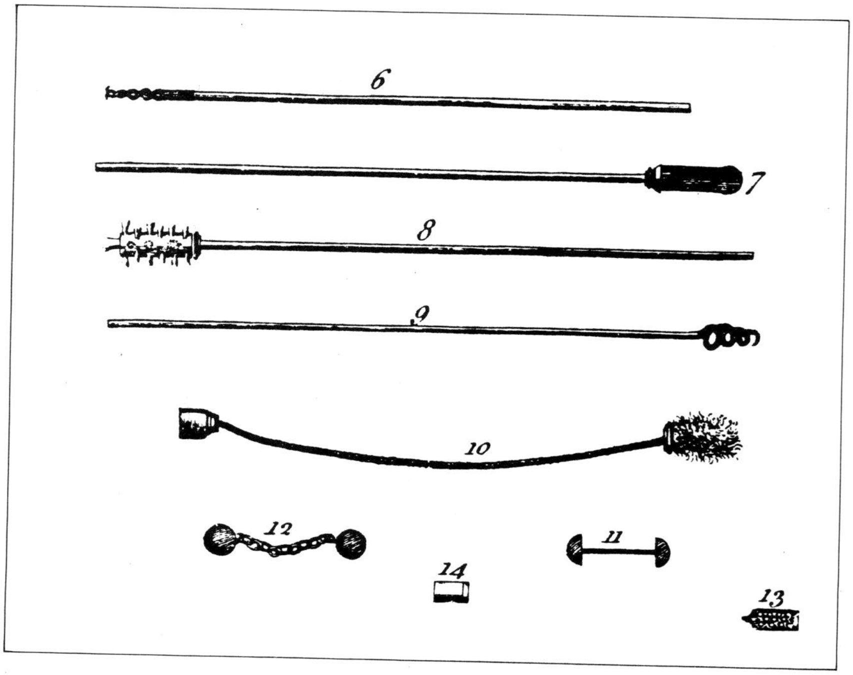

Gunnery implements and types of shot.

6. a worm, for removing a charge
7. a ladle
8. a sponge
9. another type of worm
10 a 'flexible rammer', with a rammer head at one end of a thick rope, and a sponge at the other.
11. bar shot
12. chain shot
13. grape shot
14. canister shot
(From Falconer's Marine Dictionary*)*

A selection of gunnery implements recovered from the Invincible. ►

pounder), and ran through a single block, attached to the side of the ship, and a double block, attached to the carriage. The tackles were used to haul out the gun for firing, once it had been loaded. To a certain extent they allowed the gun to be aimed for if the pull was stronger on one than the other, the angle of the gun could be roughly controlled. The third tackle was the 'train tackle' which was similar to a gun tackle but was fitted between the rear end of the carriage and a ringbolt in the centre of the deck, just outside the line of coamings. It served to hold the gun away from the port when the ship was heeling. It is possible that it was not considered a separate tackle in the mid-eighteenth century for according to a set of gunnery orders of this period, one of the gun tackles could be used as a train tackle when needed.[3]

When the gun was not in use it was 'housed', particularly in heavy weather. The gunport was closed, the quoin removed to give the gun its maximum elevation, and the whole thing run out so that its muzzle rested against the side of the ship, just above the gunport. The gun tackles were pulled tight, and 'frapped' together with the breech tackle – so that the gun was immobilised. On the lower deck, the muzzle of the gun was tied to an eyebolt situated just above the port. By an order of 1757, single bolts in this position were to be replaced by two eyebolts, side by side.

Gunpowder

Gunpowder was, of course, the most difficult of the ship's provisions to store and handle. It was made at mills at Faversham and Dartford in Kent, and at Waltham Abbey in Essex. The Faversham mill was once privately owned but was taken over by the government in 1760, just after the *Valiant* was fitted out. Powder was composed of a mixture of saltpetre, sulphur and charcoal, ground down and mixed together; some of the buildings used for this work at Faversham still survive. It was put in barrels with wooden or copper hoops to prevent sparks, carefully marked with the contents and then taken by barge to the Ordnance depot which, in this case, was Upnor Castle, across the river from Chatham Dockyard. From there it was issued to ships in the Medway. On 16 October, the *Valiant* was sent 353 barrels of gunpowder, 1,412 copper hoops for the barrels, and 11 tanned hides for the gunners' stores.[4]

Gunpowder was stored in the main magazine, forward in the hold. This area had a deck to keep it above the bilge water and partitions to keep it apart from the damp sides of the ship. The floor was constructed using a system of palleting, with drawers filled with charcoal which absorbed moisture. Forward of the magazine, on a slightly higher level, was the filling room. Here powder could be made up into cartridges of paper or canvas, and stored on racks. Entry to the magazine was only by a hatch in the roof. It was lit from the 'light room', which contained a lantern shining through a glass partition,[5] and which was totally separate from the magazine, being

entered only by a different hatch to reduce the risk of fire. There was a smaller filling room aft in the hold, known as the after powder room and used to supply the after guns in action.

Shot

Shot was rather easier to stow, the only real problem being its weight. For long-term storage it was placed in lockers in the hold, situated forward and aft of the pump well. This position, near the centre of gravity of the ship, was well chosen, as it prevented the trim of the ship being altered by any sudden expenditure of shot in action. For ready use, shot was stored in racks round the sides of the decks. After 1780 it was placed in holes in the coamings of hatchways.

Shot came in several types. Round shot was by far the most common, and a 74-gun ship of this period carried 100 rounds per gun for foreign service, and 70 rounds in home waters. In addition, it carried 'double headed' shot, shaped like a barbell and used against enemy rigging (three rounds per gun) and grape shot, consisting of musket balls formed into a cylinder inside a canvas bag, for use against personnel (five rounds per gun).[5] Other types of shot were also used on occasion such as chain shot, in which two balls, or two half balls, were joined by a chain and used against rigging; case shot, which consisted of musket balls in a tin box instead of a canvas bag; and langrel shot, consisting of iron bars. According to some authorities this was never used by the Royal Navy, but examples have recently been found aboard the wreck of the *Invincible*.

The Ordnance Board also supplied the implements needed for handling, loading and maintaining the guns. Some rammers with wooden handles were provided, with a sponge on the other end to neutralise any burning powder that might be left over from a previous firing. Rather more common then was flexible rammer and sponge which was made of thick rope, making it easier to insert in the limited space between the gun and the side of the ship. Gunners were issued with 'wadhooks' to remove charges from guns; with ladles, mostly for making up cartridges; 'aprons of lead', which covered the touch hole to protect a loaded gun from accidental firing; 'tompions', turned on a lathe and used to close the muzzle of the gun against sea and rain; and a whole range of tools for regular maintenance. The gunners' store room, forward on the orlop deck, was quite a large space but it was filled to the brim on a long voyage.

16

Manning the Ship

Captain William Brett took command of the *Valiant* on 22 August 1759. He had evidently joined the Navy as a volunteer in 1727, and been appointed captain of a guardship at Chatham in 1753. He was not a man of any great distinction and earlier in 1759, as captain of the frigate *Falmouth*, he had been court-martialled for failing to protect merchant ships being convoyed between Denmark and England. He had been found not guilty, mainly because the main complainants had failed to turn up. Brett was clearly regarded as a temporary captain for the *Valiant*; he was to fit her out and take her round to Plymouth where Captain Keppel would exchange out of the old and decrepit 74-gun ship *Torbay*.

The most immediate duty of the captain was to find a crew. Commissioned and warrant officers and midshipmen would be sent down by the Admiralty but they were useless unless enough skilled seamen could be found to man the ship. The *Valiant* had a large complement – 650 men – and the ports and harbours of Kent had already been scoured for seamen. Clearly Brett was going to have a difficult job.

The Press Gang

The main means of recruitment was the notorious press gang. A gang of ten or twelve men, headed by a lieutenant armed with a press warrant from the Board of Admiralty, could be sent ashore to grab seamen from taverns, streets or even their homes. Contrary to popular myth, however, the press gang was only allowed to take seamen. It had neither the right nor the wish to carry away landsmen for captains had no desire to fill their ships with unskilled men. Sometimes landsmen were taken up during a 'hot press' on the outbreak of war, but the great majority were soon released on proving their identity. There is no sign that Brett sent gangs ashore to Chatham and the Medway towns to find men. Any seamen who remained in the area would have been well hidden, and Brett had few men who could be spared from fitting the ship, or could be trusted not to desert if allowed ashore.

The first draft of seamen came on board on 1 September; sixty men lent from the *Hornet* sloop who were not intended, however, to form part of the permanent crew. Two days later sixty more men came from the yachts *Kathernine*, *Fubbs* and *Charlotte*, moored at Chatham.[1] These men carried out much of the heavy work of raising the masts and spars, with the assistance of some dockyard employees. On 19 September the *Valiant* received the first draft of sixty men for her own crew. They had been sent from the *Princess Royal*, the 'receiving ship' moored at the Nore, off the entrance to the Medway. This ship was well placed to guard the entrance to the Thames, and it regularly sent out boats to impress men from the homecoming merchant ships. It had recently had considerable success, and had taken two hundred men from the returning East India fleet. The agonies of such pressed men, often taken after having already had years away from home, without the chance to set foot on shore, is expressed by a letter written in 1776, by a man impressed aboard another receiving ship at the Nore.

'I was pressed out of my bed and likewise five of my shipmates. I have nothing to shift myself with, my clothes are all on board of Captain Arthur Helm. I have acquainted my father about my things which I hope I shall receive, if not I shall be in a bad condition. I have not received any answer, which makes me very ill at ease.'[2]

Finding a Crew

The *Valiant* continued to receive drafts of men from the *Princess Royal*. It also took on parties of marines from the barracks at Chatham. This corps had been formed in 1755, as an organisation under the Admiralty for providing soldiers to serve on board ship as part of the crew. Until then the army had been responsible for providing marines, and had done a poor job. The *Valiant*'s first marines, a party of ten officers and fifty privates, arrived on board on 19 September. Eventually, 124 marines would be carried, making up nearly a fifth of the ship's complement.

But the supply of seamen continued to be inadequate. As the ship prepared to sail from Blackstakes to the Nore, she had only 156 men on board, less than a quarter of her full crew. This included nearly forty officers, mid-

The popular image of the press gang. The man being taken is evidently a tailor, with scissors and tape hanging from his pocket. Attempts to seize men in the streets often led to battles with the populace.
(National Maritime Museum)

shipmen and their servants, and ninety-four marines. Of the petty officers and able seamen who would do the real work of sailing the ship, only twenty-seven were present. Even among such men as could be raised, there was a strong tendency to desert. Four men had swum from the ship on 23 September, and on the 17th three more deserted, stealing a boat and breaking the chain which had attached it to the ship.

The *Valiant* was now required for 'immediate service',[3] and urgent steps had to be taken. In the short term, dockyard workmen were sent on board to help take the ship down river and she arrived at the Nore on 21 October. The Admiralty sent orders to the Port Admiral at the Nore that the manning of the *Valiant* was to take priority, and that she was to have an elite crew chosen from the pressed men aboard the receiving ship. Two thirds of her complement were to be able and ordinary seamen and only one third, including marines, were to be landsmen. In other words, she was to have a far higher proportion of experienced men than was normal at the height of a war.[4] By the 28th the ship was fully manned, with 649 men mustered. These included nine men sick on shore, and six 'widows men' – fictitious characters whose wages were paid into a fund for the widows of officers, so only 634 men were actually on board. There were 172 petty officers and able seamen, 229 ordinary seamen and only 72 landsmen, so the order to give her an experienced crew had been carried out.

Officers

The *Valiant* carried five lieutenants, holding commissions from the Admiralty. The First Lieutenant was the most senior, and he was responsible for the discipline and organisation of the crew. The others served as watchkeeping officers. and in battle each would command a section of the ship's guns. The lieutenants lived in the wardroom, aft on the upper deck and each had a small cabin of his own, separated from the wardroom only by a canvas partition. They shared the wardroom with the Marine officers: a captain and two subalterns in a ship this size – and with the more senior warrant officers, appointed by Navy Board warrant rather than Admiralty commission. The master was an experienced seaman, responsible for the navigation of the ship, in the broadest sense. The surgeon was the head of the medical department, with three assistants under him. The purser was the chief supply officer of the ship, though in a sense he operated more as a subcontractor than as a naval officer, and he was quite entitled to make a profit on his dealings. A chaplain also formed part of the establishment of a ship of the line, but the post offered few privileges and remained unfilled on most ships.

Under the wardroom officers, and inferior to them in social status, were the 'standing officers'. They were so called because they remained with the ship even when she was out of commission, and were only transferred in the event of promotion to a higher rated ship, or the demise of their present ship (though the *Valiant* was to prove an exception to the rule). All three of the standing officers were responsible for maintenance: the gunner for the guns and their tackle, stores and ammunition; the carpenter for the hull of the ship and all her wooden fittings, including masts and boats; and the boatswain for the rigging and also for the discipline of the crew. The gunner controlled a large area aft on the lower deck, known as the gunroom though in practice most of this space was used as accommodation for junior officers. The boatswain and carpenter had small cabins on the orlop deck.

A naval lieutenant of 1777. (National Maritime Museum)

Midshipmen and Mates

Midshipmen were young men under training to become officers. In theory every midshipman had already served three years at sea, perhaps in the capacity of 'captain's servant'. In practice, the system was open to abuse and often young men could be borne on the books of a ship for some years without actually serving. The midshipmen lived together in the cockpit, aft on the orlop deck. They were often given particular responsibilities such as taking charge of one of the ship's boats, or supervising a party of seamen. They were expected to learn navigation under the master, and to become familiar with all aspects of seamanship and ship organisation. After three years as a midshipman a candidate was eligible to sit the examination for lieutenant before three captains. If he passed he was eligible for a commission, though this depended on a suitable vacancy being available. Though quite low in the hierarchy midshipmen had some of the social privileges of officers in that they were allowed to take their recreation on the quarterdeck.

All the warrant officers had mates to help them. A master's mates were educated young men often awaiting a commission as lieutenant. Under them were the quartermasters, who were petty officers in charge of the steering of the ship. The surgeon's mates were of equivalent status to the master's mate and were trained surgeons. The carpenter had a mate as second in command of his department and eight 'carpenters crew' who were also skilled craftsmen. The gunner had two mates and two 'yeomen' in charge of the magazines and storerooms. He also had eighteen 'quarter gunners', one for every four guns. There were six boatswain's mates, and these were the most prominent of the petty officers. They wielded the cat of nine tails at floggings, roused the crew from their hammocks in the mornings, and encouraged them at their work by means of beatings and shouts. The purser was assisted by a steward and his mate.

The Crew

Another part of the crew was made up of skilled artisans, such as the cooper, armourer and sailmaker. These men also had warrants from the Navy Board, but did not have the status of officers. The cook was a warrant officer too, and he was usually an old seaman who had lost a limb in the service. In practice most of the work was done by his assistants. 'Servants' formed quite a large part of the ship's company, and again their status varied. A captain's servants were mostly candidates for the midshipmen's berth, though two or three might actually look after the captain's welfare being servants in the more common sense. Each of the wardroom officers was entitled to a servant of his own, and he usually chose him from among the ship's boys or marines. In addition, a small number of servants cooked and cleaned the wardroom, as others did for the gunroom and midshipmen's berths. Some servants, such as the carpenter's, were really apprentices learning a trade.

The rest of the ship's company consisted of three categories. The able seamen had spent at least two years at sea and were skilled in all aspects of the trade, including steering, using the sounding lead and sailmaking. Ordinary seamen had spent less time at sea, but nevertheless had some skills. Landsmen were adults who had no experience of the sea, and were totally unskilled. Apart from the boys, they had the lowest status of anyone on board.

A midshipman of 1777. In the background is a ship's longboat under sail.
(National Maritime Museum)

The seamen were divided into groups for working purposes. Normally there were five or six 'parts of ship', with teams of men responsible for different tasks. Each mast had a team, known respectively as the fore, main and mizzen topmen (though on some ships the mizzen topmen were merged with the afterguard). These were young, fit men, able to run up the rigging and help take in sail in the worst of weather. They were mostly able or ordinary seamen, perhaps with a few promising landsmen attached for training. The forecastle men tended to be older, but still skilled men, for their duties included much work with the anchor and its gear. The afterguard was of lower status, though its men worked on the poop and quarterdeck under the eyes of the officers, and smartness was regarded as an asset. The lowest of all were the waisters, who worked in the waist of the ship, and were useful only for hauling on ropes, cleaning decks and the most menial work. Each part of the ship was headed by a captain for each watch, though this man had no increase of pay over an ordinary seaman, and did not even enjoy the status of petty officer. In addition to the men attached to the parts of the ship were the 'idlers', so called, not from habitual laziness, but because they did not have to stand watch throughout the day and

A seaman of the 1740s says goodbye to his girlfriend aboard ship. A hammock and a wooden tankard can also be seen. (National Maritime Museum)

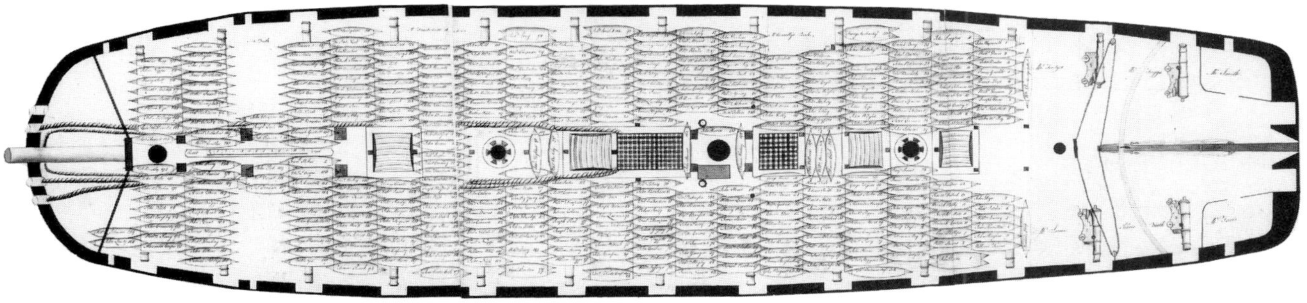

The hammock plan of a ship of around 1760 showing the gross overcrowding below decks which was accepted as normal. (The National Maritime Museum)

Artefacts from the Invincible *used for eating and drinking.*
a. Wine bottle
b. Wine bottle
c. Wooden bowl
d. Wooden bowl,.inscribed with the letters HI
e. Stoneware jar
f. Wooden tankard, with lid
g. Square plate
h. Wooden bowl, with a crudely drawn gallows carved on the base
i. Pewter bowl
j. A candle holder, possibly from an officer's cabin
k. Pewter spoon
l. Pewter plate
m. A wooden spoon, with Xs marked along the handle

night at sea. This group included the purser's steward, the cook and his mates, tradesmen such as the sail-maker and armourer, and many of the servants.

Life on Board

Up until the mid-eighteenth century, officers had taken little regard for the welfare of their men and had allowed them to run their own affairs, provided they appeared on watch on time and obeyed orders. By 1759, however, the health and welfare of crews was causing increasing concern to the officers, and their habitual uncleanliness was attracting attention. One way to counteract this was by the system of 'divisions'. The seamen were divided into small groups for welfare purposes, under a heirarchy of midshipmen and lieutenants. The divisional officers supervised the health and welfare of their men and reported to the captain. Such a system had been introduced by Admiral Smith in the Downs in 1755. Keppel was certainly well acquainted with the most advanced methods of crew organisation, for he had recently completed a four-month spell aboard the *Torbay* suffering hardly a sick man simply because of his rigorous methods of supervision. It would be very surprising if the same methods had not been pursued aboard the *Valiant*.

Off duty, the seamen had more freedom. They were allowed to form themselves into groups of eight or ten men, called 'messes', for meals. Each mess sat round a table slung between a pair of guns and the 'cook of the mess', appointed by rota, collected the food from the purser's steward, took it to the galley and then retrieved it from the cook after its preparation. The mess tables, along with the benches and stools, were made by the ship's carpenter and was yet another of the duties that had to be done as the ship was fitted out. For sleeping accommodation, the crew slung their hammocks on the lower deck, under the beams of the upper deck. Fourteen inches width was allowed per man and the situation was only mitigated by the fact that nearly half were on duty at any given time when the ship was at sea. In the daytime, the hammocks were taken up and stowed in racks round the decks and the more progressive captains, like Keppel, made sure they were washed regularly.

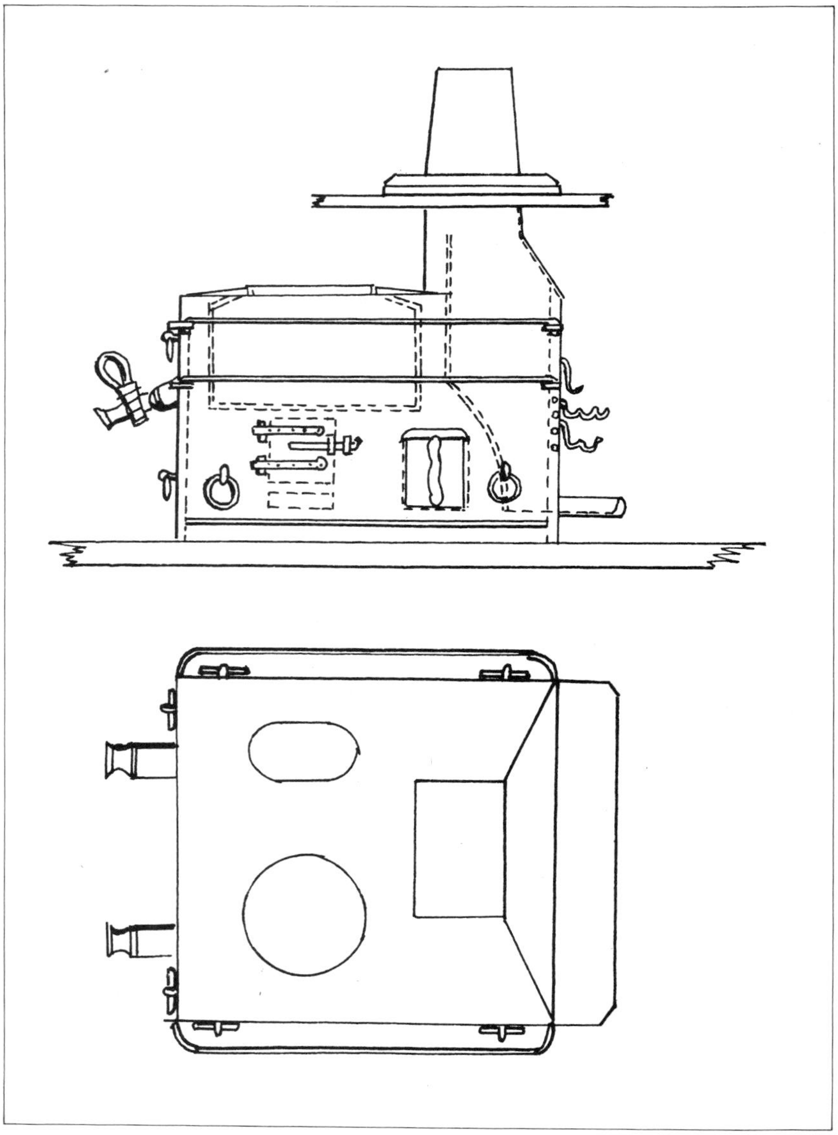

A cooking stove of 1757

Food and Drink

The diet of the seaman was basic: beef, pork, cheese, peas and a kind of hard bread known as ship's biscuit. There was a serious lack of fresh vegetables in the diet though individual captains and pursers made up for this in their purchases of provisions. For drink, the seaman mostly consumed beer. Rum was only issued in the West Indies during this period, and water was usually unpleasant after weeks at sea. Spirits were issued after the beer had run out, usually after about two months at sea.

The ship's provisions were stowed in the hold. This was the responsibility of the master, as stowage could seriously affect the trim of the ship. At the bottom level of the hold was iron ballast, consisting of pigs of iron laid on each side of the keelson. Above that was gravel ballast, which was arranged to provide a level surface in the hold. Next came the ground tier of casks, mostly filled with water or beer. (Stronger drink was kept in a separate compartment, to prevent theft.) Two more tiers of casks were placed above the ground tier which contained perishables such as salt beef, pork and cheese. The biscuit was kept in sacks, in a compartment aft of the main hold.

The *Valiant* began to take on provisions from the Victualling Yard at Chatham while she was moored in the Medway. However, these supplies were only enough for the small crew on board while she was fitting out; her full rations were sent to her at the Nore from the main victualling base at Deptford, on the Thames. The casks were brought alongside in lighters or barges, and lowered one by one into the hold through the fore, main and after hatches. When she sailed to the Nore from Blackstakes she carried 25 tons of iron ballast and 285 tons of shingle; 200 butts of beer, 38 tons of coal, wood and candles; 49 tons of round shot, 5¼ tons of grape shot and 2½ tons of double-headed shot. At the Nore she took on more provisions and would have carried approximately 10 tons each of beef and pork, 32 tons of bread and quantities of oatmeal, peas, suet and flour.[5]

The food was prepared by the ship's cook in the galley under the forecastle. The stove was made of iron and had a fireplace and a chimney. The ship's cook used the two boilers in the after side of the stove, and most of the crew's provisions were prepared in these. The fore side of the stove had a grill and a spit, with ovens on the side; these provided slightly more sophisticated cooking facilities for the officers' cooks.

Clothing

Only officers wore uniform so in theory the seaman was free to wear what he chose. In practice, his choice was limited, for he often had to buy his clothes from the purser, in the form of 'slops'. These were stored in a room on the orlop deck, and consisted of wide trousers, loose jackets, check shirts and other items and so after a period at sea the crew often began to take on a more uniform appearance.

Life at Sea

The seaman had few opportunities for recreation. Drunkenness was common, for the men were issued with eight pints of beer a day, and singing and dancing were popular, especially when a fiddler was available to play a tune. Shore leave was very limited, because of the fear of desertion, and was solely at the discretion of the captain. However, Keppel gave leave to his whole crew early in 1759, shortly before they transferred to the *Valiant*.

On duty at sea, the hours were strictly regulated. The crew (apart from idlers) was divided into two watches, starboard and larboard, and these alternated on duty. At night the work was often quite light, and only helmsmen and lookouts were needed in steady weather. At other times the men were kept constantly busy with sail trimming or with taking in sail on the approach of bad weather. The decks were cleaned daily in fair weather and the hammocks were scrubbed from time to time. Gunnery exercises were carried out as often as possible, and the sleep of the seamen was often disturbed by the need to reef down the sails, or by the approach of a suspected enemy. In the latter case, every man repaired to his 'quarters', the station allotted to him in battle. The partitions of the gundeck were taken down, the guns cast loose and the surgeon began to prepare his bandages and implements in the cockpit. The great majority of men were organised in gun crews, with fourteen men to a pair of 32-pounders. The men of the *Valiant* saw many false alarms over the years, but they also participated in several fleet battles and amphibious operations.

17

The *Valiant* in Service

By 28 November 1759 the *Valiant* was at last ready for service: fully manned, rigged, armed and stored. She was ordered to sail for Plymouth as soon as possible but remained windbound by north-easterlies for several days, unable to get her anchor up. She finally sailed on 2 November in company with the sloop *Speedwell*. The ships reached the Downs, off the east coast of Kent, and remained there for more than a fortnight, waiting for a favourable wind to take them down Channel. The *Valiant* sailed on the 17th in company with two small warships and a convoy of sixty merchantmen. The voyage to Plymouth was far from uneventful. The ships were hit by gales and the *Valiant* sprung her bowsprit. When she arrived in Plymouth Sound on Christmas Day 1759, her crew were very sickly; furthermore, no less than twenty-three men had died during Brett's short tenure of command.[1]

In the meantime, the strategic situation had changed. On 20 November, Sir Edward Hawke's fleet had chased a French fleet into Quiberon Bay in south Brittany. The French ships had been destroyed, captured or driven up the Vilaine river, to a position from which they would find it very difficult to escape. The British navy now had command of the seas, and the threat of invasion was removed. Among the ships in the thick of the action was the 74-gun *Torbay*, under Captain Keppel.

The *Torbay* was a kind of hybrid ship, a two-decker 74 cut down from an old three-decker 90, originally built in 1730 as the *Neptune*. Keppel had complained about her condition and earlier in the year he had written, 'I should be glad Lord Anson was reminded about the poor *Torbay's* bottom, as I do not hear her sheathing is ordered to be stripped off, which I am anxious of'. On another occasion, he found her performance compared unfavourably with other ships. 'The wind drawing forward, the *Torbay* did but keep her own with the chase, and then I think the *Medway* and *Dunkirk* gained on the *Torbay*.'[2] There can be little doubt that Keppel had set his heart on the new *Valiant* and used his influence with Anson to become her commander. On 11 January 1760 the *Torbay* left the scene of battle at Quiberon Bay carrying Admiral Hawke home in triumph. The ship arrived in Plymouth Sound on the 16th and arrangements began to effect the transfer. Such was Keppel's influence in high places that he was able to take his old crew with him from the *Torbay*, including the standing officers who normally stayed with their ship through all eventualities. On 28 January the crews changed ship, and Keppel took command. Brett was appointed to the *Torbay* 'until some other captain be appointed', but his service in the ship lasted for two years, so he must have been more efficient than was anticipated. The *Valiant* was still in poor condition after the gales, and was sent round to Plymouth to be docked and repaired. All the guns, stores, rigging and fittings which had been put in so painstakingly a few weeks ago were taken out, and she was put in dry dock. She was not ready for sea again until April.

Belle Ile

The *Valiant* sailed at the beginning of May, carrying £40,000 of wages to Plymouth. She sailed from that port on the 10th of the month in pursuit of a French privateer which had been spotted off the coast. She arrived in Quiberon Bay on the 26th, to join Admiral Boscawen's squadron blockading the French. She spent the rest of the year with the Channel Fleet, patrolling the French ports and lying off the Spanish port of Ferrol.

In view of the lack of activity by the French, it was decided to take some offensive action. Keppel was promoted to Commodore, with his flag in the *Valiant*. His place as captain was taken by Adam Duncan, a Scots officer who was later to become the victor of the Battle of Camperdown against the Dutch in 1797. An expedition was planned against the French island of Belle Ile, which protected Quiberon Bay against southwesterly winds, and could provide a useful base to cover the French ports of L'Orient, Port Louis and St Nazaire. Keppel was to command the expedition, the *Valiant* was to be the flagship, and the force was to consist of eleven ships of the line, thirteen smaller warships and more than ninety transports and support vessels. Eleven battalions of infantry were chosen, along with marines and artillery. The force sailed from St Helens on 29 March 1761.

Unfortunately, the French had had plenty of warning of the attack and had reinforced the defences.

The attacking force anchored off Belle Ile on 7 April, and the main attack began on the following day. It was repulsed with heavy losses (including more than fifty killed, wounded and missing from the *Valiant*), and that night gales further damaged the fleet. More troops were sent out to join the force, and another attack took place on 22 April. Artillery was landed, including some guns from the *Valiant*, and the French fortresses were bombarded; they surrendered on 22 June. As well as giving the *Valiant* her first taste of action, the affair had allowed the new marine forces to distinguish themselves; their behaviour was highly commended by Keppel and this was endorsed by the Lords of the Admiralty.

Havana

The *Valiant* remained at Belle Ile until 27 December, and then sailed to Portsmouth for docking and repair. She had been chosen for an even larger amphibious operation, still with Keppel as commodore, though he was not to be in overall command. War broke out with Spain early in January 1762 and the Cabinet decided to send an expedition against Havana, the capital of Cuba and Spain's main naval base in the Caribbean. The force was assembled at Spithead with remarkable speed and efficiency. The naval forces were commanded by Sir George Pocock and the armies by Lieutenant General the Earl of Albemarle, a relative of Keppel. When it sailed for Spithead on 5 March, it consisted of six warships and sixty-four transports. The force reached the West Indies on 27 April, but for a time the expedition was postponed because of a fear that the enemy might attack Jamaica. The force was much enlarged by other troops and ships in the West Indies, and by the time it arrived off Havana early in June, it consisted of 30 warships and 156 transports, carrying 11,000 troops.

Augustus Keppel in 1759. (National Maritime Museum)

On 7 June a group of fourteen warships, led by Keppel in the *Valiant*, covered the landing of troops to the east of Havana. The lessons of Belle Ile had been learned well and the operation was carried out with great efficiency. The troops had been carefully disposed in the transports in their regimental organisations and a well worked out code of signal flags was used to control the landing. The soldiers were put into wide, flat-bottomed boats and rowed to the shore. The Spanish were taken completely by surprise and the 11,800 troops were landed without loss.

After this initial success the campaign began to falter. The city of Havana was heavily fortified and could not be taken easily. Despite experience in the past, the British high command had not appreciated the problems caused by disease in the West Indies and with the onset of the siege soldiers began to succumb in large numbers. A naval attack was mounted on the fort of El Morro on the north side of Havana harbour but was repulsed; it was finally captured by the army despite heavy casualties. The city of

The taking of Havana. The Valiant *is the ship on the right, with bow towards the artist. (National Maritime Museum)*

Havana surrendered on 14 August and twelve ships of the line, nearly one-third of the Spanish battlefleet, were captured, along with large amounts of treasure.

The *Valiant* finally left Havana on 11 October, but she remained in Caribbean waters for another 18 months, mostly based at Jamaica. The war with France and Spain ended in 1763, but it was May 1764 before she left Jamaica for the last time in this war. She crossed the Atlantic and entered Portsmouth Harbour in July. Her officers and crew were paid off, and there must have been scenes of jubilation as the sailors were finally released from naval discipline. The *Valiant* herself was docked at Portsmouth and then laid up as one of the ships 'in ordinary'.

Her life at Portsmouth was uneventful, her only excursions being to the dry dock where she was occasionally sent for repair. In 1768, she emerged after ten days, having had a 'small repair', but in 1771 serious decay was found in her structure and it was decided to give her a 'large repair'. Eventually she was to emerge from dry dock in 1775, having had £36,297 spent on her.[3] In many ways it would have been easier and cheaper to scrap her and build a new ship, and the fact that she was expensively overhauled was a tribute to her original design, for such extensive repairs were only carried out on successful designs such as the *Victory, Invincible* and *Valiant*.

The American War

By the time *Valiant* was ready, the next war was about to start. The American colonists had initiated their revolt against British taxes, and subsequently British rule. In the

The engagement at Mona Passage, with the Valiant *attacking two enemy ships.*
(National Maritime Museum)

The Valiant *cruising in the Bay of Biscay.*
(National Maritime Museum)

early stages of the war only the smaller ships were needed, for the Americans had no navy of their own; but as war with France, Spain, and eventually Holland, became increasingly probable, ships like the *Valiant* began to be made ready for service. She was commissioned on 11 November 1776 under Captain Leveson Gower, an associate of Augustus Keppel's. The latter officer had risen in rank since the last war, and in March 1778 he was made Commander-in-Chief of the fleet in home waters. The *Valiant*, after some months cruising against American shipping in the Bay of Biscay, was ordered to join his force. War with France began in July and the *Valiant* was part of Keppel's fleet sent to blockade the main French base at Brest.

The French navy had been revived since its defeats in the Seven Years War. New ships had been built, and this time France had no continental enemy to distract her. On 23 July the lookout at the masthead of the *Valiant* sighted the French fleet coming out to sea. The information was signalled to the rest of the fleet, and the ships cleared for action. At last, nearly 20 years after her launch, the *Valiant* was ready to carry out the task for which she had been designed – to fight the French in a great fleet battle. The ensuing action proved to be a disappointment. The fleets remained in sight of one another for four days, but exchanged no fire. Then, on 27 July a number of British ships had their top hamper damaged before the French escaped under cover of night. Keppel claimed that he had not been fully supported by his second-in-command, Sir Hugh Palliser. A long series of courts-martial ensued which caused dissension and disunity among the officer corps and Leveson Gower resigned from the command of the *Valiant* in March 1779, partly in support of Keppel.

The new captain was Samuel Goodall. The ship remained in the home fleet under successive admirals, and the war intensified with the entry of Spain on the French side. On 31 August, under Sir Charles Hardy, the fleet was forced to retreat from a combined French and Spanish fleet which outnumbered them. For a time the enemy was left in control of the English Channel, but failed to exploit the opportunity.

In March 1781 the *Valiant* formed part of the escort of a large convoy sent to relieve Gibraltar, which was beseiged by the Spanish. She took part in several cruises on the Western approaches to the Channel, but in the following year was ordered to the West Indies, where the main strength of the navy was now being concentrated. On 9 April the fleet, under Admiral Rodney, was passing south of the French island of Guadeloupe when the *Valiant*'s lookout again sighted the enemy fleet. Yet again the fleets were in contact for several days, with a short engagement on the 9th, followed by a full-scale battle on the 12th. The two fleets passed one another in line ahead, when Rodney made the daring and unprecedented move of cutting the enemy line with his flagship. It led to the defeat of the French, the capture of six enemy ships of the line, and heralded the tactics which were to be used by Nelson in later battles. The *Valiant* played a full part in the action, as part of the red division of the fleet. She had ten men killed and twenty-eight wounded. A few days later, she played a leading role in the capture of two French stragglers from the action, the *Cato* and the *Jason*.

The war ended in the following year, with the British recognising the independence of the American colonies, and losing Minorca and Florida to Spain. The *Valiant* was returned home and was paid off in June 1783 at Plymouth. Again she was laid up in ordinary, but in 1790 there was the threat of war with Spain over Nookta Sound, near modern Vancouver. The fleet was mobilised, and the *Valiant* was commissioned under the Duke of Clarence. This officer was a nephew of the King and in 1830 he was to become King himself, as William IV, the Sailor King. The *Valiant* took part in some exercises with the fleet, but the dispute with Spain was settled and she was paid off at Plymouth in November. It was to be the Duke's last command of a seagoing ship for he was promoted Rear Admiral soon afterwards.

The Duke of Clarence around the time when he commanded the Valiant.

The French Revolutionary War

The French Revolution had already begun by this time, and relations with the new rulers of France were deteriorating. War with France began in 1793, but it was April of the following year before the *Valiant*, now regarded as an old ship, was able to put to sea, under the command of Captain Thomas Pringle. She joined the Channel Fleet under Lord Howe, and almost immediately she was involved in the great battle known as the Glorious First of June. The French, protecting a large grain convoy which was

The Glorious First of June, 1794. The scene after the battle. (National Maritime Museum)

Stangate Creek, an engraving after Turner, 1826. It is possible that one of the ships moored in the left background is the Valiant.

essential to the survival of the revolutionary government, were first sighted on 28 May. The main battle took place three days later, and again the *Valiant* was fully engaged, having two men killed and nine wounded. The French lost six ships of the line, though the grain convoy escaped.

The *Valiant* remained in the Channel Fleet for two more years, mostly escorting important convoys. In 1796 she was again sent to the West Indies, though that theatre was far less important than it had been in the last war. Finally in 1799, under Captain John Crawley, she was sent home to England. In April she returned to the Medway for the first time since her building there forty years earlier. She was surveyed and found to be in poor condition. In September, the officers of Chatham Yard reported that the *Conquerant*, previously intended for use as a lazaretto, or quarantine ship, in the Medway was 'in too bad a condition to be employed in that service'. They recommended that the *Valiant*, no longer fit for seagoing service, should be used for that role.[4] She was converted to a static harbour ship and taken to Stangate Creek near Queenborough where she remained for more than a quarter of a century. She was finally broken up at Sheerness in 1826, and it is possible that some of her timbers were re-used in work in that yard, or even at Chatham; perhaps some of them survive today.

The Development of the 74

The 74-gun ship remained the standard ship of the line throughout the life of the *Valiant*. The *Triumph* and *Valiant* were highly successful examples, but the policy of copying from the French was not pursued for much longer. Probably Anson believed in developing the skills of native British designers, and he found his favourite shipwright in Sir Thomas Slade who, as Surveyor of the Navy, designed most of the early British 74s. The Slade draughts – for the *Bellona, Arrogant, Ramillies, Culloden* and *Albion* classes – remained in standard use even after his death in 1771. He was succeeded by Sir John Williams who designed some 74s of his own though he never matched Slade in skill. By the later stages of the American War, however, the old type of 74, usually 168ft long on the gundeck, was regarded as too small. French ships were back in fashion and the navy had found a new ship to copy from – the *Courageaux*, taken by the *Bellona* in 1761. She was copied many times from 1779 onwards, either directly or with modifications. There was a tendency to build ever larger ships over the next fifteen years, culminating in 1795. In this climate the *Invincible* design found favour again, and was revived for two ships, the *Kent* and *Ajax*. However, ideas on size had moved even further by this time, and a centre section was added to each ship during building, to bring the length up to 182ft. This exceeded the safe limits for a ship of two decks, and the keel of the *Kent* was later found to have arched 18in. After that there was a tendency towards slightly smaller ships, about 176ft on the gundeck. In 1806 a new class was designed by the two Surveyors of the Navy, using a variant of the lines of the *Courageaux*; this class, later known as the 'Forty thieves' to its detractors, was a disappointment in service.

The whole picture changed in 1811 when Robert Seppings, Master Shipwright at Chatham and later Surveyor of the Navy, developed a new system of construction which allowed much longer ships. Two-deckers of 80- or even 90-guns became possible, and by the 1820s the 74 was obsolescent. Until then no ship was built which was a significant improvement on the *Valiant* of 1759. In 1802 her sister the *Triumph* was described as 'a ship which, for the space of more than twenty years, was considered the finest of her class then existing'.[5] Only the fact that the *Valiant* had been withdrawn from active service a few years earlier prevented her from earning the same accolade.

The name continues to be used by the Royal Navy. Another 74, of the Surveyors Class, was launched in 1807 and broken up in 1823, while the original ship was still in existence in the Medway. The third ship was an iron armoured ship, built in 1863. She had a long existence after being reduced to harbour service in 1888, and was not broken up until 1956. The fourth *Valiant* was completed in 1914, as one of the mighty *Queen Elizabeth* class battleships. She fought at the Battle of Jutland in 1916, against the Italians at Matapan in 1941, and took part in several operations in the Mediterranean and Far East. She was broken up in 1950. The fifth *Valiant* was launched in 1963, the first nuclear submarine with a British reactor. She is still in service with the navy.

Chatham Dockyard was used by the Royal Navy until 1984, and opened to the public in the following year as Chatham Historic Dockyard. It still has many of the buildings which were in use when the *Valiant* was built, along with exhibitions and demonstrations of crafts such as sail, flag and rope-making. The building of the first *Valiant* is commemorated today in the Wooden Walls exhibition. It tells the story of the building of the ship, through the eyes of John North, her carpenter, and William Crockwell, his apprentice. It is sited in the old Masthouses and mould loft, where the moulds for the *Valiant* were made and her masts constructed.

Notes

Chapter 1

1. Fox, Frank *Great Ships* (London 1980) p59
2. Lavery, B *Ship of the Line,* vol 1 (London 1983) p90
3. *Ibid,* p90
4. *Ibid,* pp91–3
5. *Ibid,* p107

Chapter 2

1. Lavery, B *The Royal Navy's First Invincible* (Portsmouth 1987) p3
2. Archives National, Paris, Marine G 246. National Maritime Museum, Greenwich, SPB/34
3. PRO, Adm 3/57, 8 August 1747
4. *The Royal Navy's First Invincible,* pp71 & 73
5. PRO, Adm 95/12, 21 May 1757

Chapter 3

1. Franklin, John *Navy Board Ship Models* (London 1989) p177

Chapter 4

1. Hasted, Edward *Historical and Topographical Survey of the County of Kent* (first published 1797–1801, reprinted Canterbury 1972) vol IV, p192
2. Information supplied by Mrs S Horley of Loughton, Essex
3. Indenture of William Wilkins, NMM SPB/20
4. Navy Records Society, Naval Administration 1715–50, edited by D. Baugh, 1977, pp300–1
5. PRO Adm 106, 6 October 1758
6. PRO Adm 106 2188, ff350–2, 362–4, 372
7. Pool, B *Navy Board Contracts* (London 1966) pp94–5

Chapter 5

1. PRO Adm 106/2199, 9 August 1768
2. House of Commons Report, 1771, p14
3. PRO Adm 106/2199, 9 August 1768
4. Charnock J *A History of Marine Architecture,* (London 1800–2), vol III, pp140–57
5. House of Commons Report, 1771, p4
6. PRO Adm 196/2189, 2 March 1757
7. House of Commons Report, 1771, p64, appendix IV
8. Quoted in Jonathan Coad, *The Royal Dockyards, 1690–1850* (London 1989) p127
9. *Ibid*

Chapter 6

1. Quoted in R A Salaman *Dictionary of Tools used in the Woodworking Trades, c1700 to 1790* (London 1975) p29
2. Falconer, W *Marine Dictionary* (1769, reprinted Newton Abbott 1970) p123
3. *Ibid,* p298

Chapter 7

1. Falconer, p170
2. NMM, POR/J/1

Chapter 8

1. Falconer, p242
2. Navy Records Society, The Tomlinson Papers, 1835, p226

Chapter 9

1. Falconer, p282
2. Steel, D *The Shipwright's Vade Mecum,* (Second edition, London 1822) p130

Chapter 10

1. NMM, POR/J/1

Chapter 11

1. Falconer, p60
2. Lavery, B *The Arming and Fitting of English Ships of War* 1600–1815 (London 1987) p58
3. Falconer, p213
4. *The Mariner's Mirror,* vol 70, 1984, p329

Chapter 12

1. Burney, W *Universal Dictionary of the Marine,* 1815, (Reprinted London 1970) p xiv

Chapter 13

1. *The Mariner's Mirror,* vol 63, 1976, pp137–44

Chapter 14

1. Roke, Margaret and Wyman, John *Essays in Kentish History,* (London 1973) (edited), 'A tour into Kent', p187

Chapter 15

1. Navy Records Society, The Sandwich Papers, vol II, 1933, p168
2. Lavery, B *The Arming and Fitting of English Ships of War 1600–1815* p275
3. *Ibid,* p141
4. PRO Adm 160/2
5. Lavery, B *The Arming and Fitting of English Ships of War 1600–1815* p288

Chapter 16

1. PRO Adm 51/3998, log of the *Valiant*
2. From a private collection, by courtesy of Mr C Pickering
3. PRO Adm 1/716
4. *Ibid,* 23 October 1759
5. PRO Adm 51/3998

Chapter 17

1. PRO Adm 1/802, 25 December 1759. PRO Adm 51/3998
2. PRO Adm 1/2010, 28 May 1758
3. PRO Adm 180/2, Admiralty Progress Books, p89
4. PRO Adm 2/289, 13 September 1799
5. Charnock, J *History of Marine Architecture,* (London 1800–2) Vol III, p144

PRO Public Records Office, Kew
NMM National Maritime Museum, Greenwich

Chronology

1757

11 January
Navy Board orders two ships to the draught of the *Dublin* class, one to be built at Chatham.

31 March
Chatham dockyard ordered to send estimate to the Navy Board. New ship to be named *Valiant*.

21 May
Admiralty orders *Triumph* and *Valiant* to be built to the draught of the *Invincible*.

21 July
Draught for the *Valiant* sent to Chatham.

2 August
Chatham dockyard ordered to send new estimate for the *Valiant*.

19 August
New estimate sent to the Navy Board: £33,170 for the hull, masts and yards, and £7,172 for rigging and stores.

7 September
Dimensions of scantlings sent by Navy Board to Chatham.

1758

1 February
Work begun on the *Valiant*.

29 March
Work on the *Valiant* 'in hand', and estimated to cost £6,000 in the current year.

1759

2 March
Estimate for work on the *Valiant* in the current year – £13,778. Completion estimated for March 1760.

4 April
Dimensions of masts and yards for the *Valiant* prepared; these to be ready by the time the ship is launched.

18 April
Deptford Yard ordered to send two anchors of 69 and 72cwt to Chatham for the *Valiant*.

21 May
Riggers ordered to be taken off the *Sandwich* and put on the *Valiant* instead.

31 July
Master Shipwright at Chatham announces that the works of the *Valiant* 'are in such forwardness that he purposes to launch her the ensuing spring tides', and the Navy Board gives orders for the launch.

6 August
Valiant graved in the dock

10 August
Valiant floated out of the dock.

22 August
At the moorings off Princes' Bridge, Captain William Brett takes command.

19 September
First pressed men received from *Princess Royal*.

10 October
Sailed to Blackstakes. Guns put in board.

16 October
Powder received from Upnor Castle.

21 October
Arrived at the Nore.

28 October
Valiant ready for service.

6 November
Valiant sails to the Downs.

Bibliography

Manuscript Sources

1. Public Record Office, Admiralty records

Adm 1, Admiralty in letters
Adm 2, Admiralty out letters, especially Lords' Letters, which include the main orders relating to shipbuilding
Adm 95/12, orders for shipbuilding, 1719–62
Adm 42 series, Dockyard muster books
Adm 51 series, captains' logs
Adm 106 series, Navy Board papers, including in letters from the dockyards, letters to the Admiralty, etc.
Adm 180 series, especially 180/2. Progress Books, giving basic information on the building and repairs of ships, including costs.

2. National Maritime Museum

Draught Room: numerous draughts of ships, including those of the *Valiant*, and of many other ships of the same period, showing details of construction.

Manuscripts: ADM/B series, Navy Board to Admiralty letters
CHA/E series, Navy Board orders to Chatham dockyard
POR/J/1 and 2. Weekly reports on work done to ships at Portsmouth

Contemporary Works on Naval Architecture and Related Subjects

Blanckley, T, *The Naval Expositor*; 1750, reprinted Rotherfield, East Sussex, 1988
Dodd, R, *Days at the Factories*; 1843
Falconer, W, *Marine Dictionary*; first published 1768, reprinted, Newton Abbot, 1970
Fincham, J, *Introductory Outline of the Practice of Shipbuilding*; 1821
Murray, M, *A Treatise on Shipbuilding and Navigation*; 1765
Panoucke, *Enclyclopedie Methodique; Marine;* 1783–7, reprinted Nice, 1986–7
Rees, A, *The Cyclopeadia*; 1819–20, part reprinted Newton Abbot, 1970 under the title *Naval Architecture.*
Stalkaart, M, *Naval Architecture*; 1781
Steel, D, *Elements and Practice of Naval Architecture*; 1805, reprinted London 1977
The Shipwright's Vade Mecum; 2nd edition, 1822
The Elements and Practice of Rigging and Seamanship; two vols, 1794, reprinted in London, 1978

Other Printed Primary Sources

Baugh, D, ed Naval Administration, 1715–1750; Navy Records Society, 1977
House of Commons, Report on the Supply of Timber; 1771
Knight, R, Portsmouth Dockyard Papers, 1774–1783; 1987

Later Works on Shipbuilding, etc

Albion, RG, *Forests and Sea Power*; Cambridge, Mass 1926
Boudriot, J, *The 74-Gun Ship*; 4 vols, London, 1986–8
Bugler, AR, *HMS Victory, Building, Restoration and Repair*; London, 1966
Dodds, J and Moore, J, *Building the Wooden Fighting Ship*; London 1984
Goodwin, P, *The Construction and Fitting of the Sailing Man of War*; London, 1987
Lavery, B, *The Ship of the Line*; 2 vols, London, 1983–4
Anatomy of the Ship – The 74-Gun Ship Bellona; London, 1985
The Arming and Fitting of English Ships of War: 1600–1815; London, 1987
The Royal Navy's First Invincible; London, 1988
Longridge, CN, *The Anatomy of Nelson's Ships*; London, 1955
Salaman, RA, *Dictionary of Tools Used in the Woodworking and Allied Trade, c1700–1970*; London, 1975

Dockyards

Coad, J, *The Historic Architecture of the Royal Navy*; London, 1983
The Royal Dockyards, 1690–1850; Aldershot, 1989
McDougall, P, *The Chatham Dockyard Story*; Rochester, 1981
Morriss, R, *The Royal Dockyards During the Revolutionary and Napoleonic Wars*; Leicester, 1983
Pool, B, *Navy Board Contracts*; London, 1966

General and Naval History

Baugh, D, *British Naval Administration in the Age of Walpole*; Princeton, 1966
Rodger, NAM, *The Wooden World*; London, 1986
Schomberg, Capt I, *Naval Chronology*; vol 5, 1802
Syrett, D, *The Seige and Capture of Havana*; Navy Records Society, 1970

Index

Page numbers in italic indicate that the reference is to an illustration. An asterisk after a page number indicates the main section on the subject.